Digital Electronics

AUTHOR

Dr. Menka Yadav
Assistant Professor,
Dept. of Electronics and Communication Engineering (ECE),
Malviya National Institute of Technology,
Jaipur-302017, Rajasthan

REVIEWER

Dr. Priyadarsan Parida
Associate Professor,
Dept. of Electronics and Communication Engineering (ECE),
GIET University , School of Engineering and Technology,
Gunupur, Rayagada-765022, Orissa

KHANNA BOOK PUBLISHING CO. (P) Ltd.
4C/4344, Ansari Road, Darya Ganj,
New Delhi, India

All India Council for Technical Education
Nelson Mandela Marg, Vasant Kunj,
New Delhi, India

Dear Readers,

To prevent the piracy, this book is secured with HIGH SECURITY HOLOGRAM on the front title cover. In case you don't find the hologram on the front cover title, please write us to at contact@khannabooks.com or whatsapp us at +91-99109 09320 and avail special gift voucher for yourself.

Specimen of Hologram on front Cover title:

Moreover, there is a SPECIAL DISCOUNT COUPON for you with EVERY HOLOGRAM.

How to avail this SPECIAL DISCOUNT:

Step 1: Scratch the hologram

Step 2: Under the scratch area, your "coupon code" is available

Step 3: Logon to www.khannabooks.com

Step 4: Use your "coupon code" in the shopping cart and get your copy at a special discount

Step 5: Enjoy your reading!

ISBN: 978-93-5538-729-5

Book Code: DIP439EN

Digital Electronics

by Dr. Menka Yadav

First Edition: 2025

Originally Published by:
All India Council for Technical Education, Ministry of Education, GOI

Exclusively Published & Distributed by:
Khanna Book Publishing Co. (P) Ltd.
4C/4344, Ansari Road, Daryaganj,
New Delhi (INDIA)

Visit us at: **www.khannabooks.com**
Write us at: **contact@khannabooks.com**

To view complete list of books, Please scan the QR Code:

Printed in Bharat.

AICTE'S LIST OF PRESCRIBED TEXTBOOKS

As Per Model Curriculum & Outcome Based Education

aligned with

NATIONAL EDUCATION POLICY 2020

UNDERGRADUATE COURSE BOOKS

S.No.	ISBN	Book Name	Author
1	9789355382160	Advanced Electrical Machines: Theory and Practicals	Bharat Singh Rajpurohit
2	9789355382139	Analog Electronic Circuits: Theory and Practical	Venu Sonti, Sachin Jain
3	9789355383808	Analog Electronic Devices: Theory and Practicals	N. B Balamurugan
4	9789391505158	Basic Electrical Engineering (with Lab Manual)	S.K. Sahdev
5	9789355388087	Basic of Electronics	Brijesh Iyer
6	9789355387127	Basics of Electrical Machines: Theory and Practicals	D. K. Palwalia, U. K. Kalla, R. K. Kumawat
7	9789355389381	Basics of Mechanics	Shubhankar Bhowmick, N. Venkata Swamy Naidu
8	9789355386250	Basics of Thermodynamics	Parmod Kumar, Atul Dhar
9	9789355386984	Biology for Engineers	Annamma Odaneth
10	9789391505141	Chemistry-I	Manisha Agarwal
11	9789355384843	Civil Engineering Societal and Global Impact	Shakuntala Acharya
12	9789355382184	Computer Aided Civil Engineering Drawing	D. Chandra Mouli, B. Swathi
13	9789355382580	Computer Aided Design & Analysis	Hari Kumar Voruganti
14	9789355387592	Computer Architecture	D. Vaithiyanathan
15	9789355385093	Computer Networks	Hari Prabhat Gupta
16	9789355385307	Construction Engineering & Management	Sparsh Johari
17	9789355388728	Control Systems	S.N. Deepa
18	9789355385420	Data Structure and Algorithms	Ranbir Singh Sanasam
19	9789355386700	Design and Analysis of Algorithms	Hari P. Gupta, Rahul Mishra
20	9789355388872	Digital Electronics and System	Abhishek Bhatt
21	9789355386571	Digital Signal Processing	Sanjeev Sharma
22	9789355383396	Disaster management	Pradeep K. Goyal, Anil K. Gupta
23	9789355385222	Discrete Mathematical Structures	Narendra S. Chaudhari, Sharmila S. P.
24	9789355387868	Effective Technical Communication	Pritee Sharma, Sanchita Verma
25	9789355383389	Electrical Circuit Analysis and Network Theory	Shouri Chatterjee
26	9789355388957	Electromagnetic Waves	Uttam M. Pal, Srijith K.
27	9789355384652	Electronic Communication	Vimal Bhatia
28	9789355388711	Embedded Systems	Santanu Chattopadhyay
29	9789355385147	Energy Science and Engineering	M. Rizwan, Majid Jamil
30	9789355387929	Engineering Economics, Estimation & Costing	Yogesh C. Joshi
31	9789355384669	Engineering Geology	R. N. Khare, Ketan Chourasia
32	9789391505066	Engineering Graphics & Design	Pradeep Jain
33	9789391505097	English (with Lab Manual)	Kulbhushan Kumar
34	9789355385963	Environmental Engineering	Chandra Sekhar Matli
35	9789355388117	Fluid Mechanics	Shreenivas N. Londhe, Pradnya R. Dixit
36	9789355388377	Fluid Mechanics & Fluid Machines	R.P. Saini
37	9789355387103	Geotechnical Engineering	Paramita Bhattacharya

UNDERGRADUATE COURSE BOOKS

S.No.	ISBN	Book Name	Author
38	9789355385130	Hydraulic Engineering	Sayantan Ganguly
39	9789355385550	Hydrology & Water Resources Engineering	Mitthan Lal Kansal, P.K. Agarwal
40	9789355389688	Information Technology Workshop in MATLAB	Hari Prabhat Gupta
41	9789355383235	Instrumentation for Civil Engineering Applications	Ashim Kanti Dey, Manita Das
42	9789355387042	Introduction to Civil Engineering	Manabendra Saharia, Nagendra R. Velaga
43	9789355384805	Introduction to Database Systems	Vikram Singh
44	9789355387509	Machine Element & System Design	Sabah Khan
45	9789355385383	Machine Learning	Himanshu Mittal
46	9789355388414	Manufacturing Automation	Chitragupt Swaroop Chitransh
47	9789355387233	Manufacturing Processes	Dheerendra Kumar Dwivedi
48	9789355389886	Materials Engineering	Dheerendra Kumar Dwivedi
49	9789391505172	Mathematics - I Calculus and Linear Algebra [For Non-Computer Science Engineering Branches]	Reena Garg
50	9789391505271	Mathematics - I Calculus and Linear Algebra [For Computer Science Engineering Branches]	Reena Garg
51	9789355384607	Measurements & Metrology	V.P. Chandramohan
52	9789355383730	Mechanics of Materials	Arvind Vyavahare
53	9789355383358	Mechatronics, Robotics & Control	J. Srinivas
54	9789355385116	Microprocessor and Microcontroller	Saurabh Chaudhury, Risha Mal
55	9789355386083	Microprocessors	Yogendra Kumar Gupta
56	9789355389169	Mobile Communication and Networks	Vimal Bhatia
57	9789355385475	Operating Systems	Sukomal Pal
58	9789355385604	Operating Systems	Sanjaya Kumar Panda
59	9789391505134	Physics (Oscillations, Waves and Optics) With lab manual	A.B. Bhattacharya, Atanu Nag
60	9789391505257	Physics (Quantum Mechanics for Engineers) With lab manual	A.B. Bhattacharya, Atanu Nag
61	9789391505165	Physics (Introduction to Electromagnetic Theory) (with Lab Manual)	A.B. Bhattacharya, Atanu Nag
62	9789391505059	Physics (Introduction to Mechanics) (with Lab Manual)	A.B. Bhattacharya, Atanu Nag
63	9789355385956	Power Electronics: Theory and Practicals	Lalit Chandra Saikia
64	9789355384942	Power Systems-I	Suresh Mikkili
65	9789355386236	Power Systems-II	Arun Rahul S, Manas K. Jena
66	9789355383754	Product Innovation & Entrepreneurship	Vigneswaran C
67	9789355384584	Production & Operation Management	Navin Kumar Dev
68	9789391505219	Programming for Problem Solving (with Lab Manual)	R.S. Salaria
69	9789355385451	Signals and Systems	Sanjay L. Nalbalwar
70	9789355386021	Strength of Materials	Uday Shanker Dixit, Nelson Muthu, S. M. Kamal
71	9789355383792	Structural Engineering	T.D. Gunneswara Rao, D. Ravi Prasad
72	9789355387004	Surveying and Geomatics	P. K. Garg
73	9789355384317	Testing & Evaluation of Civil Engineering Materials	Sandeep Chaudhary, Kashyap A Patel
74	9789355382306	Theory of Computation	Debi Prasanna Acharjya
75	9789355384928	Transportation Engineering	A.K. Raji
76	9789355385208	VLSI Design	Mahendra Sakare, Shonal Chouksey
77	9789391505332	Workshop / Manufacturing Practices (with Lab Manual)	Veeranna D.K.

DIPLOMA COURSE BOOKS

S.No.	ISBN	Book Name	Author
1	9789355382481	Aanandam An Act of Community Service	Shuchi Sharma
2	9789355385567	Advanced Manufacturing Processes	Ranjeet Kumar Sahu, Devendra Laxman Kamble
3	9789355385109	Advanced Surveying: Theory & Practice	Ramakant Agrawal, Parshottam Sarathe
4	9789355384546	Algorithms	Piyoosh P, Arnab Sarkar
5	9789391505448	Applied Chemistry (with Lab Manual)	Anju Rawlley, Devdatta Vinayakrao Saraf
6	9789391505431	Applied Physics-I (with Lab Manual)	Vinod Kumar Yadav, Mina Talati
7	9789391505578	Applied Physics-II (with Lab Manual)	Hussain Jeevakhan
8	9789355387530	Basic Mechanical Engineering	Dharmendra Singh, Chandra Shekhar Rajoria
9	9789355387394	Building Construction	Rinku Kumar, Sandeep Panchal
10	9789355383280	Building Electrification	Arnab Ghosh
11	9789355384379	Building Planning and Drawing	Aparna Dwivedi
12	9789391505493	Communication Skills in English	Anjana Tiwari
13	9789355383570	Computer Aided Machine Drawing Practice	Kanak Kalita
14	9789355388964	Computer Networking and Data Communication	Sanjaya Shankar Tripathy
15	9789355382986	Computer Networks: Theory & Practicals	Brijendra Pratap Singh, Manoj Madhava Gore
16	9789355382672	Computer Programming: Theory and Practicals	Satyendra Singh Chouhan
17	9789355387714	Computer System Organization	Sonal Yadav
18	9789355382658	Concrete Technology: Theory and Practice	Sasankasekhar Mandal, Vishwajit Anand, Sushil K. Agarwala
19	9789355383334	Construction Materials and Testing	Vanita Aggarwal
20	9789355383501	Consumer Electronics	Amit M. Joshi, Maulin M. Joshi, Urvashi Prakash Shukla
21	9789355387684	Data Structures: Theory & Practicals	NB Venkateswarlu
22	9789355388513	Design of Machine Elements	A. Kumaravel, M. Kathirselvam
23	9789355388360	Design of Steel and RCC Structure	Sparsh Johari
24	9789355383303	Digital Communication	Muralidhar Kulkarni, K. S. Shivaprakasha
25	9789355387295	Digital Electronics	Menka Yadav
26	9789355389435	Electric Circuits and Networks	Jamuna K., Nilanjan Tewari
27	9789355388452	Electric Motors and Transformers (Theory and Practicals)	Sanjay B. Bodkhe
28	9789355385901	Electric Power Transmission and Distribution	Sudhir Sharma
29	9789355389466	Electrical & Electronics Measurements and Instrumentation	Sudarsan Sahoo, Vipin Chandra Pal, Sudipta Chakraborty
30	9789355382696	Electronic Communication	M. D. Selvaraj, P. Nagaradjane
31	9789355387202	Electronic Devices and Circuits	Ankesh Jain
32	9789355386373	Electronic Equipment Maintenance	Chanchal Sharma
33	9789355384690	Embedded Systems	Deepa M, G. Santhanamari, Susithra. N
34	9789355384867	Energy Conservation and Audit	M. Rizwan, Majid Jamil
35	9789391505509	Engineering Graphics	Sharad K. Pradhan, K.K. Jain
36	9789391505608	Engineering Mechanics (with Lab Manual)	Bhankhar Bharat, Gokaldas Vandana Somkuwar

DIPLOMA COURSE BOOKS

S.No.	ISBN	Book Name	Author
37	9789391505516	Engineering Workshop Practice	A. K. Sarathe
38	9789391505653	Environmental Science	Subrat Roy
39	9789355384966	Estimating, Costing and valuation	Sandeep Panchal
40	9789355389510	Fluid Mechanics & Hydraulic Machinery	Suman Chakraborty, Sourav Mitra, Aditya Bandopadhyay
41	9789391505592	Fundamentals of Electrical and Electronics Engineering (with Lab Manual)	Susan S. Mathew, Saji T. Chacko
42	9789355387240	Fundamentals of Power Electronics	Soumitra Kumar Mandal
43	9789355389602	Geotechnical Engineering (Theory & Practicals)	Neelima Satyam
44	9789355386137	Induction, Synchronous, and Special Electrical Machines	K. Selvajyothi, S. Tamilselvi, Sarat Kumar Sahoo
45	9789355382290	Industrial Engineering & Management	A. Manimaran
46	9789355384959	Internet of Things	Sujata Pal
47	9789355388650	Introduction to DBMS: Theory & Practicals	Myneni Madhu Bala
48	9789355384294	Introduction to e-Governance	Kamal Kant Verma
49	9789355386564	Introduction to Electric Generation Systems	Lalit Chandra Saikia, Nalin Behari, Deb Choudhury
50	9789391505585	Introduction to IT Systems (with Lab Manual)	Prashant Joshi
51	9789355382900	Linear Integrated Circuit	B. T. Krishna
52	9789355388148	Manufacturing Engineering	Santosh Kumar
53	9789355383877	Material Science & Engineering	Vinod Kumar, Nisheeth k. Prasad
54	9789391505424	Mathematics-I	Deepak Singh
55	9789391505523	Mathematics-II	Garima Singh
56	9789355385994	Measurements and Metrology	A. Gnanavelbabu, M. Santhana kumar, R. Saranraj
57	9789355388131	Mechanics of Materials and their Testing	R. Velmurugan, Pragathi
58	9789355385048	Microcontroller Applications	Ritula Thakur
59	9789355385062	Microcontrollers and Applications	Santanu Chattopadhyay
60	9789355388100	Mobile and Wireless Communication	Maulin Joshi, Urvashi P. Shukla
61	9789355387790	Production & Operations Management	Mayank Kumar Agrawal
62	9789355388704	Public Health Engineering	Rinku Kumar
63	9789355387455	Python Programming	Rupesh Nasre
64	9789355387189	Renewable Energy Technologies	Sanjay Agrawal, Rajeev K. Mishra
65	9789355387479	Software Engineering	Salman Abdul Moiz
66	9789355389749	Theory of Machines & Mechanisms	G. C. Mohan Kumar
67	9789355387332	Theory of Structures	Arunachalam Subramanian Balu
68	9789355389015	Thermal Engineering - I	V. Raghavan
69	9789355389831	Thermal Engineering - II	Parvesh Antil
70	9789355386779	Transportation Engineering (Theory & Practice)	A.K. Raji, K.K. Babu
71	9789355384287	Water Resource Engineering (Theory & Practice)	Prakash C. Swain, Dillip K. Ghose
72	9789355383051	Web Technologies: Theory & Practical	Raman Dugyala

प्रो. म. जगदीश कुमार
अध्यक्ष
Prof. M. Jagadesh Kumar
Chairman

सत्यमेव जयते

आज़ादी का अमृत महोत्सव

अखिल भारतीय तकनीकी शिक्षा परिषद
(भारत सरकार का एक सांविधिक निकाय)
(शिक्षा मंत्रालय, भारत सरकार)
नेल्सन मंडेला मार्ग, वसंत कुंज, नई दिल्ली–110070
दूरभाष : 011–26131498
ई–मेल : chairman@aicte-india.org

ALL INDIA COUNCIL FOR TECHNICAL EDUCATION
(A STATUTORY BODY OF THE GOVT. OF INDIA)
(Ministry of Education, Govt. of India)
Nelson Mandela Marg, Vasant Kunj, New Delhi-110070
Phone : 011-26131498
E-mail : chairman@aicte-india.org

FOREWORD

Engineers are the backbone of the modern society. It is through them that engineering marvels have happened and improved quality of life across the world. They have driven humanity towards greater heights in a more evolved and unprecedented manner.

The All India Council for Technical Education (AICTE), led from the front and assisted students, faculty & institutions in every possible manner towards the strengthening of the technical education in the country. AICTE is always working towards promoting quality Technical Education to make India a modern developed nation with the integration of modern knowledge & traditional knowledge for the welfare of mankind.

An array of initiatives have been taken by AICTE in last decade which have been accelerate now by the National Education Policy (NEP) 2020. The implementation of NEP under the visionary leadership of Hon'ble Prime Minister of India envisages the provision for education in regional languages to all, thereby ensuring that every graduate becomes competent enough and is in a position to contribute towards the national growth and development through innovation & entrepreneurship.

One of the spheres where AICTE had been relentlessly working since 2021-22 is providing high quality books prepared and translated by eminent educators in various Indian languages to its engineering students at Under Graduate & Diploma level. For the second year students, AICTE has identified 87 books at Under Graduate and Diploma Level courses, for translation in 12 Indian languages - Hindi, Tamil, Gujarati, Odia, Bengali, Kannada, Urdu, Punjabi, Telugu, Marathi, Assamese & Malayalam. In addition to the English medium, the 1044 books in different Indian Languages are going to support to engineering students to learn in their mother tongue. Currently, there are 39 institutions in 11 states offering courses in Indian languages in 7 disciplines like Biomedical Engineering, Civil Engineering, Computer Science & Engineering, Electrical Engineering, Electronics & Communication Engineering, Information Technology Engineering & Mechanical Engineering, Architecture, and Interior Designing. This will become possible due to active involvement and support of universities/institutions in different states.

On behalf of AICTE, I express sincere gratitude to all distinguished authors, reviewers and translators from different IITs, NITs and other institutions for their admirable contribution in a very short span of time.

AICTE is confident that these out comes based books with their rich content will help technical students master the subjects with factor comprehension and greater ease.

(Prof. M. Jagadesh Kumar)

ACKNOWLEDGEMENT

The authors are grateful to the authorities of AICTE, particularly Prof. M. Jagadesh Kumar, Chairman; Prof. M. P. Poonia, Vice-Chairman; Prof. Rajive Kumar, Member-Secretary and Dr Amit Kumar Srivastava, Director, Faculty Development Cell for their planning to publish the books on Digital Electronics. We sincerely acknowledge the valuable contributions of the reviewer of the book Dr. Priyadarsan Parida, Associate Professor, Department of Electronics and Communication Engineering, School of Engineering and Technology, GIET University, Gunupur, Rayagada-765022, Orissa, for making it students' friendly and giving a better shape in an artistic manner.

I convey thanks to my PhD scholars Ms Shalini Chaudhary, Ms Basudha Dewan and Mr. Devender Pal Singh for their valuable contribution to help me to bring this book in its final shape within the given time.

This book is an outcome of various suggestions of AICTE members, experts and authors who shared their opinion and thought to further develop the engineering education in our country. Acknowledgements are due to the contributors and different workers in this field whose published books, review articles, papers, photographs, footnotes, references and other valuable information enriched us at the time of writing the book.

Dr. Menka Yadav

PREFACE

The book titled "Digital Electronics" is an outcome of the rich experience of our teaching of Digital Electronics course to Engineering students. The initiation of writing this book is to expose Digital Electronics concepts to Diploma students, the fundamentals of Digital Electronics as well as enable them to get an insight of the subject. Keeping in mind the purpose of wide coverage as well as to provide essential supplementary information, we have included the topics recommended by AICTE, in a very systematic and orderly manner throughout the book. Efforts have been made to explain the fundamental concepts of the subject in the simplest possible way.

During the process of preparation of the manuscript, we have considered the various standard text books and accordingly we have developed sections like critical questions, solved and supplementary problems etc. While preparing the different sections emphasis has also been laid on definitions and laws and also on comprehensive synopsis of formulae for a quick revision of the basic principles. The book covers all types of medium and advanced level problems and these have been presented in a very logical and systematic manner. The gradations of those problems have been tested over many years of teaching to a wide variety of students.

Apart from illustrations and examples as required, we have enriched the book with numerous solved problems in every unit for proper understanding of the related topics. It is important to note that we have included the relevant laboratory practical. In addition, besides some essential information for the users under the heading "Know More" we have clarified some essential basic information in the appendix and annexure section.

As far as the present book is concerned, "Digital Electronics" is meant to provide a thorough grounding in digital electronics concepts and its applications to various combinational, sequential logic circuits, memory and data converters. This book will prepare diploma students to apply the knowledge of digital electronics to tackle 21st century and onward engineering challenges and address the related aroused questions. The subject matters are presented in a constructive manner so that diploma students are prepared to work in different sectors or in national laboratories at the very forefront of technology.

We sincerely hope that the book will inspire the students to learn and discuss the ideas behind basic principles of digital electronics and will surely contribute to the development of a solid foundation of the subject. We would be thankful to all beneficial comments and suggestions which will contribute to the improvement of the future editions of the book. It gives us immense pleasure to place this book in the hands of the teachers and students. It was indeed a big pleasure to work on different aspects covering in the book.

Dr. Menka Yadav

OUTCOME BASED EDUCATION

For the implementation of an outcome based education the first requirement is to develop an outcome based curriculum and incorporate an outcome based assessment in the education system. By going through outcome based assessments evaluators will be able to evaluate whether the students have achieved the outlined standard, specific and measurable outcomes. With the proper incorporation of outcome based education there will be a definite commitment to achieve a minimum standard for all learners without giving up at any level. At the end of the programme running with the aid of outcome based education, a student will be able to arrive at the following outcomes:

PO-1: Basic and discipline specific knowledge: Apply knowledge of basic mathematics, science and engineering fundamentals and engineering specialization to solve the engineering problems.

PO-2: Problem analysis: Identify and analyse well-defined engineering problems using codified standard methods.

PO-3: Design/development of solutions: Design solutions for well-defined technical problems and assist with the design of systems components or processes to meet specified needs.

PO-4: Engineering Tools, Experimentation and Testing: Apply modern engineering tools and appropriate technique to conduct standard tests and measurements.

PO-5: Engineering practices for society, sustainability and environment: Apply appropriate technology in context of society, sustainability, environment and ethical practices.

PO-6: Project Management: Use engineering management principles individually, as a team member or a leader to manage projects and effectively communicate about well-defined engineering activities.

PO-7: Life-long Learning: Ability to analyse individual needs and engage in updating in the context of technological changes.

COURSE OUTCOMES (CO)

Students will be able to:

CO1: Understand Number System

CO2: Interpret Boolean Logic Expressions

CO3: Design Combination Logic Circuits

CO4: Understand Sequential Logic Circuits

CO5: Compare Memory Devices based on Technology, Size and Operation

CO6: Understand Data Conversion

Correlation of COs and POs

Course Outcomes	EXPECTED MAPPING WITH PROGRAMME OUTCOMES (1- Weak Correlation; 2- Medium correlation; 3- Strong Correlation)						
	PO-1	PO-2	PO-3	PO-4	PO-5	PO-6	PO-7
CO-1	3	1	-	-	-	1	1
CO-2	3	2	2	1	1	1	2
CO-3	3	3	3	2	1	2	2
CO-4	3	3	3	1	2	2	2
CO-5	2	2	1	-	2	-	2
CO-6	3	2	2	2	3	3	3

GUIDELINES FOR TEACHERS

To implement Outcome Based Education (OBE) knowledge level and skill set of the students should be enhanced. Teachers should take a major responsibility for the proper implementation of OBE. Some of the responsibilities (not limited to) for the teachers in OBE system may be as follows:

- Within reasonable constraint, they should manipulate time to the best advantage of all students.
- They should assess the students only upon certain defined criterion without considering any other potential ineligibility to discriminate them.
- They should try to grow the learning abilities of the students to a certain level before they leave the institute.
- They should try to ensure that all the students are equipped with the quality knowledge as well as competence after they finish their education.
- They should always encourage the students to develop their ultimate performance capabilities.
- They should facilitate and encourage group work and team work to consolidate newer approach.
- *They should follow Blooms taxonomy in every part of the assessment.*

Bloom's Taxonomy

Level	Teacher should Check	Student should be able to	Possible Mode of Assessment
Creating	Students ability to create	Design or Create	Mini project
Evaluating	Students ability to Justify	Argue or Defend	Assignment
Analysing	Students ability to distinguish	Differentiate or Distinguish	Project/Lab Methodology
Applying	Students ability to use information	Operate or Demonstrate	Technical Presentation/ Demonstration
Understanding	Students ability to explain the ideas	Explain or Classify	Presentation/Seminar
Remembering	Students ability to recall (or remember)	Define or Recall	Quiz

GUIDELINES FOR STUDENTS

Students should take equal responsibility for implementing the OBE. Some of the responsibilities (not limited to) for the students in OBE system are as follows:

- Students should be well aware of each Unit Outcome (UO) before the start of a unit in each and every course.
- Students should be well aware of each Course Outcome (CO) before the start of the course.
- Students should be well aware of each Programme Outcome (PO) before the start of the programme.
- Students should think critically and reasonably with proper reflection and action.
- Learning of the students should be connected and integrated with practical and real life consequences.
- Students should be well aware of their competency at every level of OBE.

ABBREVIATIONS AND SYMBOLS

List of Abbreviations

General Terms			
Abbreviations	**Full form**	**Abbreviations**	**Full form**
CO	Course Outcome	PS	Present state
LSB	Least Significant Bit	NS	next state
MSB	Most Significant Bit	MOD	Modulus
DIP	Dual-in-pakage	SISO	Serial in serial out
PDF	Power spectral density	PISO	parallel in serial out
SR	Set Reset	SIPO	serial in parallel out
SNR	Signal to noise ratio	PIPO	parallel in parallel out
SOP	Sum of products	ADC	analog to digital converter
POS	product of sums	DAC	digital to analog converter
NM	Noise Margin	SAR	Successive approximation register
HA	Half adder	Gnd	Ground
HS	Half subtractor	S/H	Sample & Hold
FA	Full adder	FSR	Full scale range (Voltage)
FS	Full subtractor	INL	Integrated Non-linearity
K-map	Karnaugh map	DNL	Dynamic non linearity
MUX	Multiplexer	IC	Integrated circuit
DEMUX	Demultiplexer	WLP	Wafer level packaging
RAM	Random access memory	CPU	Central Programming Unit
DRAM	Dynamic RAM	FF	Flip flop
SRAM	Static RAM	EPROM	Electrically PROM
DDR-RAM	Double data rate RAM	EEPROM	Electrically Erasable ROM
DDR-SDRAM	DDR synchronous DARM	FET	Field Effect Transistor
ROM	Read only memory	MOSFET	Metal oxide semiconductor FET
PROM	Programmable ROM	CMOS	Complimentary MOS

List of Units and Symbols

Symbols	Description	Symbols	Description
$\oplus$	XOR binary operation	V_{ref}	Reference voltage
t_{su}	Setup time	C_H	Hold capacitor
t_h	hold time	σ_e	Quantization error
t_r	rise time	Δ	Step size
t_f	fall time	V_{in}	input voltage
V_{LSB}	Least significant bit voltage	V_{out}	output voltage

LIST OF FIGURES

LIST OF TABLES

CONTENTS

1 Number Systems & Boolean Algebra

UNIT SPECIFICS

Through this unit we have discussed the following aspects:

- Digital Versus Analogue
- Number Systems Introduction
- Decimal Number System
- Binary Number System
- Octal Number System
- Hexadecimal Number System
- 1's and 2's Complement Representation
- Boolean Variables Rules and Laws of Boolean Algebra
- Karnaugh Maps and its Use for Simplification of Boolean Expressions.

The practical applications of the topics are discussed for generating further curiosity and creativity as well as improving problem solving capacity.

Besides giving a large number of multiple choice questions as well as questions of short and long answer types marked in two categories following lower and higher order of Bloom's taxonomy, assignments through a number of numerical problems, a list of references and suggested readings are given in the unit so that one can go through them for practice. It is important to note that for getting more information on various topics of interest some QR codes have been provided in different sections which can be scanned for relevant supportive knowledge.

After the related practical, based on the content, there is a "Know More" section. This section has been carefully designed so that the supplementary information provided in this part becomes beneficial for the users of the book. This section mainly highlights the initial activity, examples of some interesting facts, analogy, history of the development of the subject focusing the salient observations and finding, timelines starting from the development of the concerned topics up to the recent time, applications of the subject matter for our day-to-day real life or/and industrial applications on variety of aspects, case study related to environmental, sustainability, social and ethical issues whichever applicable, and finally inquisitiveness and curiosity topics of the unit.

RATIONALE

This fundamental unit on number system helps students to get a primary idea about the different number systems and how the conversion from one number system to other number system can be done. These include several numerical representation schemes as well as binary representations of numeric and alphabetic data. This chapter go into great details about conversion between various number systems. Binary arithmetics like addition, subtraction operations are discussed.

This chapter also discusses Boolean algebra, Boolean theorems and its various postulates, and minimization of logic functions using Boolean Algebra and Karnaugh map minimization techniques.

PRE-REQUISITES

Basic understanding of digital signals.

UNIT OUTCOMES

List of outcomes of this unit is as follows:

U1-O1: Describe basic difference between digital and analog quantities

U1-O2: Describe the binary, octal, decimal, hexadecimal number system

U1-O3: Explain conversion between different number systems

U1-O4: Describe basic laws and rules of Boolean algebra

U1-O5: Describe the use of Karnaugh map to simplify Boolean expression

Unit-1 Outcome	EXPECTED MAPPING WITH COURSE OUTCOMES (1 - Weak Correlation; 2 - Medium Correlation; 3 - Strong Correlation)					
	CO-1	CO-2	CO-3	CO-4	CO-5	CO-6
U1-O1	3	3	-	-	-	-
U1-O2	3	-	-	-	-	-
U1-O3	3	2	3	-	-	-
U1-O4	-	3	3	2	-	-
U1-O5	-	2	3	2	-	-

"A friend in need is friend indeed & Knowlegde is your best friend."

Dr. Menka Yadav

1.1 ANALOG VS DIGITAL

1.1.1 Analog Signals

Numerous systems produced information-carrying signals which are continuous both in terms of quantities and timing. As digital signals have become more prevalent, analog signals are being used less and less. To put it simply, analogue signals are any signals that are generated naturally or occur naturally.

1.1.2 Digital Signals

Digital signals, in contrast to analogue signals, are discrete in both value and time. These signals, which are made up of various voltage levels, are represented by binary integers. Figure 1.1 demonstrates the Analog and Digital signals.

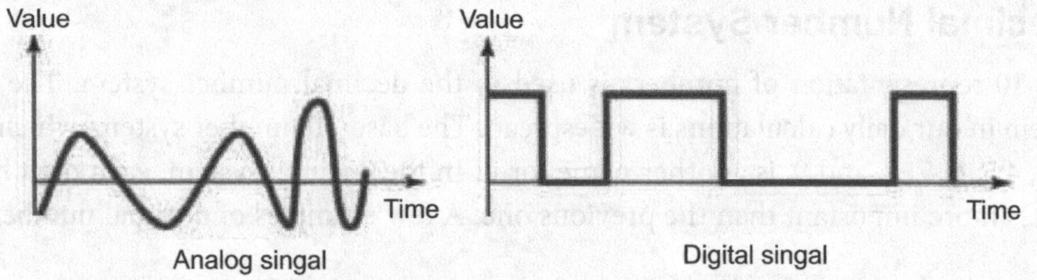

Fig. 1.1: .Analog versus Digital Signals

1.1.3 Analog versus Digital Signals

Analog Signals	Digital Signals
Continuous signals	Discrete signals
Represented by sine waves	Represented by square waves
Continuous range of values	Discontinuous valures
Records sound waves as they are	Converts into a binary waveform.
Only used in analog devices.	Suited for digital electronics like computers, mobiles and more.

1.1.4 Advantages of digital circuits over analogcircuits

(i) The accuracy of the digital circuit is excellent.
(ii) Signals encoded digitally may be transported without noise-related signal deterioration, in contrast to analog systems.
(iii) Digital circuit components are highly affordable and simple to make in numerous components on a single chip.
(iv) Digital signals are superior at maintaining quality across long distances.
(v) Compared to analog communications, digital signals may contain more information per second.
(vi) Compared to analog circuits, digital circuits are easier to construct and less expensive.
(vii) Digital signals are more rapid than analog ones.
(viii) A digital circuit's hardware implementation is more adaptable than analog one.

1.2 NUMBER SYSTEMS

A number system with base 'r' will have 'r' different digits ranging from '0' to 'r-1'. Decimal equivalent of a number can be found using below formulae from any number system. For more details scan QR code.

Scan to know more about Number System

Example: $(101.1011)_r = 1xr^2 + 0xr^1 + 1xr^0 + 1xr^{-1} + 0xr^{-2} + 1xr^{-3} + 1xr^{-4}$

r = base or radix.

Here the number from left side of decimal point is multiplied (every digit) by ri where number is located at ith position from decimal (starting from 0,1,2,3,...). The fractional part is multiplied by r-i, where 'i' is the position of the digit from the decimal point (starting from 1,2,3...). Different number systems and symbols (digitals) required are listed in the following table.

1.2.1 Decimal Number System

The base (r)-10 representation of numbers is used in the decimal number system. The usage of this number system in our daily calculations is widespread. The base-10 number system, which has 10 digits like 0, 1, 2, 3, 4.5, 6, 7, 8, and 9, is another name for it. In the decimal system, each digit has a location and is 10 times more important than the previous one. A few examples of decimal numberg system are given below-

Base	Different Digits
2-Binary	0,1
4	0,1,2,3
6	0,1,2,3,4,5
8-Octal	0,1,2,3,4,5,6,7
10-Decimal	0,1,2,3,4,5,6,7,8,9
12	0,1,2,3,4,5,6,7,8,9, A, B
16-Hexadecimal	0, ……,9, A, B, C, D, E, F

$$(92)_{10} = 9 \times 10^1 + 2 \times 10^0$$

$$(200)_{10} = 2 \times 10^2 + 0 \times 101 + 0 \times 10^0$$

$$(212.367)_{10} = 2 \times 10^2 + 1 \times 10^1 + 2 \times 10^0 + 3 \times 10^{-1} + 6 \times 10^{-2} + 7 \times 10^{-3}$$

1.2.2 Binary Number System

A number that is written with base 2 or the binary system is referred to as a binary number system. It uses the symbols 1 and 0 to represent different numeric values. The positional notation using 2 as a radix is known as the base-2 system.

Example: $(1011.1)_2 = 1 \times 23 + 0 \times 22 + 1 \times 21 + 1 \times 20 + 1 \times 2^{-1} = (11.5)_{10}$

- **Advantages of Binary Number System:**

  For usage in computer systems, the binary number system clearly outperformed previous number systems. This number system's ability to simply express all types of data in terms of 0s and 1s was yet another huge benefit. Additionally, two clearly separate modes of operation for simple electrical devices utilized for hardware implementation may be carried out quickly and effectively. When the data is represented as 0s and 1s, the circuits needed to execute mathematical operations like add, subtract, multiply, division, etc., become simple.

1.2.3 Octal Number System

The octal number system includes eight different digits since its radix is 8.There are eight independent digits: 0, 1, 2, 3, 4, 5, 6, and 7. In this number system, the place values for the various digits are 8^0, 8^1, 8^2, and so on (for the integer portion) as well as 8^{-1}, 8^{-2}, 8^{-3}, and so forth (for the fractional part).

Example: $(1011.1)_8 = 1 \times 8^3 + 0 \times 8^2 + 1 \times 8^1 + 1 \times 80 + 1 \times 8^{-1} = (521.125)_{10}$

1.2.4 Hexadecimal Number System

The 16 fundamental digits of the hexadecimal number system, which is a radix-16 number system, are 0 to 9 and A to F. For the integer portion, hexadecimal numbers start with 16^0, then 16^1, then 16^2, then 16^3, and so forth (for the fractional portion).

Example: $(1011.1)_{16} = 1 \times 16^3 + 0 \times 16^2 + 1 \times 16^1 + 1 \times 160 + 1 \times 16^{-1} = (4113.0625)_{10}$

1.3 NUMBER REPRESENTATION IN BINARY

Numbers in binary or any other number system can be represented in two ways.

A. Signed magnitude
B. Complement representation-

Complement representation is further divided in two parts

- Reduced base or r-1's complement
- Radix or r's complement

1.3.1 Sign-Bit Magnitude

The MSB represents the "sign" bit, where "0" signify a plus sign and a "1" denote a minus sign. The magnitude is represented by the remaining bits.

Example

$$(00101011)_2 = +(43)_{10}$$
$$(10101011)_2 = -(43)_{10}$$

1.3.2 1's Complement Representation

To obtain 1's complement of any binary number- replace '0' with '1'and '1' with '0' in the given binary number.

Example 1.1

Find 1's complement of 10101110.

Solution:

1's complement will be 01010001.

Example 1.2

Find 1's complement of 10001.001.

Solution:

1's complement will be 01110.110.

1.3.2.1 1's Complement for Signed Binary numbers

Positive binary numbers are represented same as it is (no change). For negative binary number, the 1's complement is used to represent it.

Example 1.3

The representation of –5 and +5 in 1's complement form

Solution:

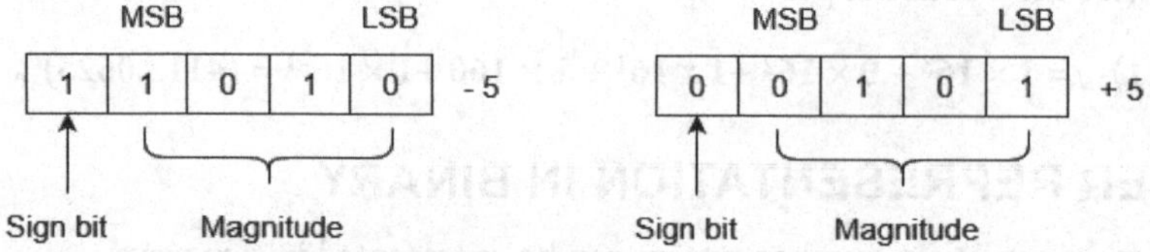

Drawback: In this system, 0 has two alternative representations:

-0 (1 1111)

+0 (0 0000)

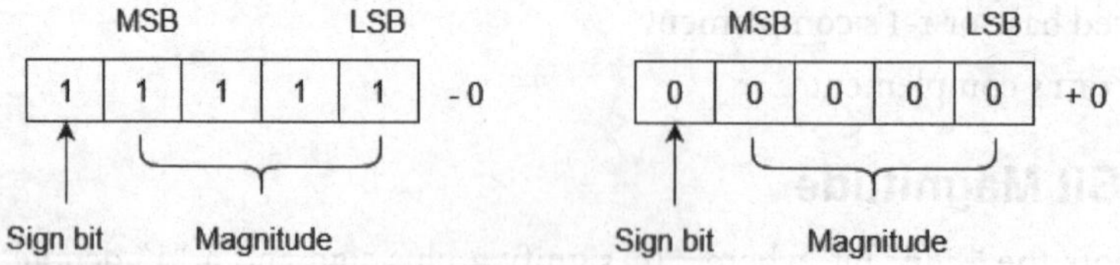

1.3.3 2's Complement Representation

Take 1's complement of the given number and add 1 to the LSB of the output to obtain the binary integer's 2's complement. Or in other words find 1s complement and add 1 in the LSB.

Example 1.4

Find 2's complement of 10101110.

Solution:

Given Number = 10101110

1's complement = 01010001

01010001

+1

= 01010010

Example 1.5

Find 2's complement of 10001.001.

Solution:

Given Number = 10001.001

1's complement = 01110.110

01110.110

+1

= 01110.111

1.3.3.1 2's Complement for Signed Binary number Representation

Positive binary numbers are represented as it is (no change) while for a negative number, the 2's complement is used to represent it.

Example 1.6

The representation of -5 and +5 in 2's complement form.

Solution:

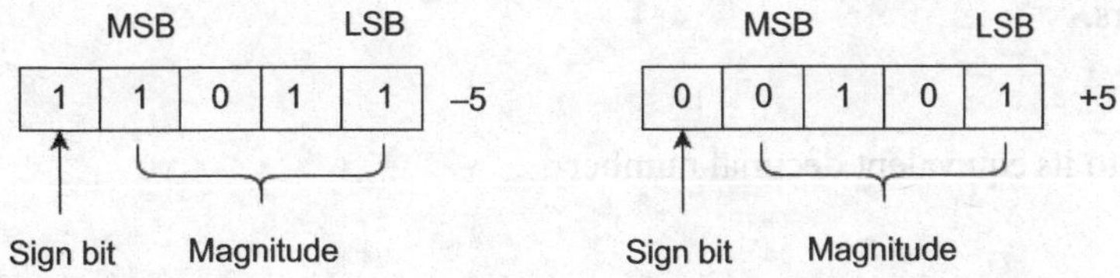

- This system's benefit is that 0 only has one representation for –0 and +0. In the 2's complement form, zero (0) is seen as always positive (sign bit is 0). It is therefore a singular portrayal.

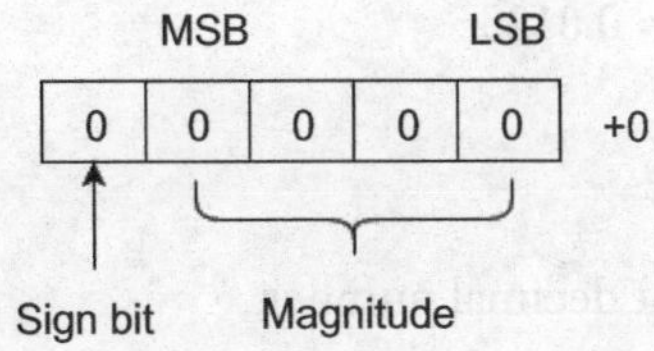

1.4 NUMBER SYSTEM CONVERSION

The need of number system conversion is there since as human beings we are more comfortable with decimal number systems. So, any other number systems are difficult to understand for us. But on the other hand, side computers/processors can understand binary number system easily but can't understand decimal number system. So, number system is creating clear interface between human beings and processors.

1.4.1 Binary-to-Decimal Conversion

- Rules for converting integer
  - (i) Each binary digit's positional value should be known. From right to left, multiply each digit by 2^{0}, 21, 22, 23, 24, and so forth.
  - (ii) Add together all of these numbers to get the corresponding decimal number.

Example 1.7

Convert $(1011010)_2$ to equivalent decimal number.

Solution:

$$(1011010)_2$$

$$= 1 \times 2^6 + 0 \times 2^5 + 1 \times 2^4 + 1 \times 2^3 + 0 \times 2^2 + 1 \times 2^1 + 0 \times 2^0$$
$$= 64 + 0 + 16 + 8 + 0 + 2 + 0$$
$$= 90$$

Therefore, $(1011010)_2 = (90)_{10}$

- Rules for converting fraction
  - (i) By multiplying each digit of a binary fraction number by 2^{-1} 2^{-2} 2^{-3} 2^{-4}, and so on, starting from the left, you may get each digit's positional value.
  - (ii) Add together all of these numbers to get the fractional binary number's equivalent in decimal numbers.

Example 1.8

Convert (0.1101)2 to its equivalent decimal number

Solution:

(0.1101)2

$$= 1 \times 2^{-1} + 1 \times 2^{-2} + 0 \times 2^{-3} + 1 \times 2^{-4}$$
$$= 0.5 + 0.25 + 0 + 0.0625$$
$$= 0.8125$$

Therefore, $(0.1101)_2 = (0.8125)_{10}$

Example 1.9

Convert $(1111111.0101)_2$ to equivalent decimal number.

Solution:

(i) Considering the Integer Part

$(1111111)_2$

$$= 1 \times 26 + 1 \times 25 + 1 \times 24 + 1 \times 23 + 1 \times 22 + 1 \times 21 + 1 \times 20$$
$$= 64 + 32 + 16 + 8 + 4 + 2 + 1$$
$$= 90$$

(ii) Considering the Fractional part

$(0.0101)_2$

$$= 0 \times 2^{-1} + 1 \times 2^{-2} + 0 \times 2^{-3} + 1 \times 2^{-4}$$
$$= 0 + 0.25 + 0 + 0.0625$$
$$= 0.3125$$

Therefore, $(1111111.0101)_2 = (90.3125)_{10}$

1.4.2 Octal-to-Decimal Conversion

- Rules for converting integer
  - (i) Each octal digit's positional value should be known. From right to left, multiply each digit by 8^0, 8^1, 8^2, 8^3, 8^4, and so forth.
  - (ii) Add together all of these numbers to get the corresponding decimal number.

Example 1.10

Convert $(16512)_8$ to equivalent decimal number.

Solution:

$$(16512)8$$

$$= 1 \times 8^4 + 6 \times 8^3 + 5 \times 8^2 + 1 \times 8^1 + 2 \times 8^0$$

$$= 4096 + 3072 + 320 + 8 + 2$$

$$= 7498$$

Therefore, $(16512)_8 = (7498)_{10}$

- Rules for converting fraction
  - (i) By multiplying each digit of a binary fraction number by 8-1 8-2 8-3 8-4, and so on, starting from the left, you may get each digit's positional value.
  - (ii) Add together all of these numbers to get the fractional binary number's equivalent in decimal numbers.

Example 1.11

Convert (0.145)8 to equivalent decimal number.

Solution:

$$(0.145)8$$

$$= 1 \times 8\text{-}1 + 4 \times 8\text{-}2 + 5 \times 8\text{-}3$$

$$= 0.125 + 0.0588 + 0.0098$$

$$= 0.1936$$

Therefore, $(0.145)_8 = (0.1936)_{10}$

1.4.3 Hexadecimal to - Decimal Conversion

- Rules for converting integer
  - (i) Each hexadecimal digit's positional value should be known. From right to left, multiply each digit by 16^0, 16^1, 16^2, 16^3, 16^4, and so forth.
  - (ii) Add together all of these numbers to get the corresponding decimal number.

Example 1.12

Conversion of (13D1A)16 to equivalent decimal number.

Solution:

$$(13D1A)_{16}$$

$$= 1 \times 16^4 + 3 \times 16^3 + 13 \times 16^2 + 1 \times 16^1 + 10 \times 16^0$$

$$= 65536 + 12288 + 3328 + 16 + 10$$

$$= 81178$$

Therefore, $(13D1A)_{16} = (81178)_{10}$

- Rules for converting fraction
    (i) By multiplying each digit of a binary fraction number by 16^{-1} 16^{-2} 16^{-3} 16^{-4}, and so on, starting from the left, you may get each digit's positional value.
    (ii) Add together all of these numbers to get the fractional binary number's equivalent in decimal numbers.

Example 1.13

Convert (0.2A5)16 to equivalent decimal number.

Solution:

(0.2A5)16

$= 2 \times 16^{-1} + 10 \times 16^{-2} + 5 \times 16^{-3}$

$= 0.125 + 0.0391 + 0.0012$

$= 0.1653$

Therefore, $(0.2A5)_{16} = (0.1653)_{10}$

Example 1.14

Conversion of $(1351A.2A5)_{16}$ to equivalent decimal number

Solution:

(i) Considering the Integer part

$(1351A)_{16}$

$= 1 \times 16^4 + 3 \times 16^3 + 5 \times 16^2 + 1 \times 16^1 + 10 \times 16^0$

$= 65536 + 12288 + 1280 + 16 + 10$

$= 79130$

(ii) Considering the Fraction part

(0.2A5)16

= 2 × 16-1 + 10 × 16-2 + 5 × 16-3

$= 0.125 + 0.0391 + 0.0012$

$= 0.1653$

Therefore, $(1351A.2A5)_{16} = (79130.1653)_{10}$

1.4.4 Decimal to Binary Conversion

- Rules for converting integer
    (i) Calculate the reminder by dividing the decimal number by the binary number's base (2)
    (ii) Divide the result by 2 once again, note down the reminder, and repeat this process until the result is 0.
    (iii) The remainder, which is the binary counterpart of the provided decimal number, should now be written down in the opposite order.

Example 1.15

Conversion of $(27)_{10}$ to equivalent binary number.

Solution:

2	27	1
2	13	1
2	6	0
2	3	1
2	1	1
	0	

Therefore, $(27)_{10} = (11011)_2$

- Rules for converting fraction
  - (i) Write the integer component you obtain after multiplying the fractional part by two. Otherwise, type 0.
  - (ii) Once more, multiply the fraction portion of the result from the first step you obtained, write down the integer part, and repeat this process until the fraction part of the result is zero (minimum 4 to 5 times if it repeats).
  - (iii) Now, list your integers or zeros in the same order as you received them.

Example 1.16

Conversion of (0.45)10 to equivalent binary number.

Solution:

Multiplication	Result	Integer Portion	Fraction Portion
0.45 × 2	0.90	0	.90
0.90 × 2	1.80	1	.80
0.80 × 2	1.60	1	.60
0.60 × 2	1.20	1	.20
0.20 × 2	0.40	0	.40

Therefore, $(0.45)_{10} = (0.\ 01110..)_2$

1.4.5 Decimal to Octal Conversion

- Rules for converting integer:
  - (i) Calculate the reminder by dividing the decimal number by the octal number's base (8).

Example 1.17

Conversion of $(127.375)_{10}$ to equivalent binary number.

Solution:

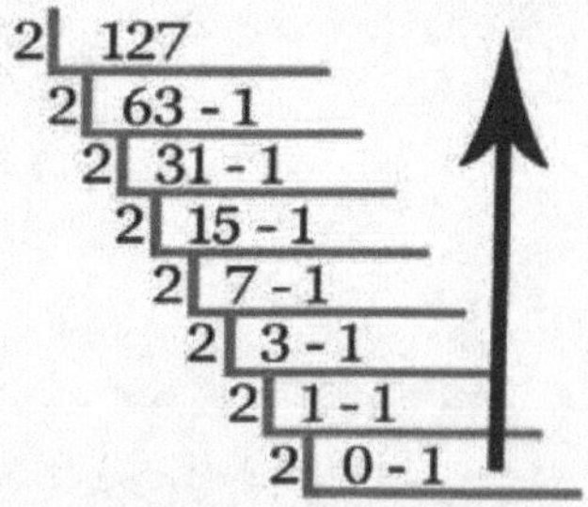

$(127)_{10} = (1111111)_2$

Multiplication	Result	Integer Portion	Fraction Portion
0.375 × 2	0.75	0	.75
0.750 × 2	1.50	1	.50
0.50 × 2	1.00	1	.00

Therefore, $(127.375)_{10} = (1111111.011)_2$

(ii) Divide the result by 8 once again, note down the reminder, and repeat this process until the result is 0.

(iii) The remainder, which is the octal counterpart of the provided decimal number, should now be written down in the opposite order.

Example 1.18

Convert $(158)_{10}$ to equivalent octal number.

Solution:

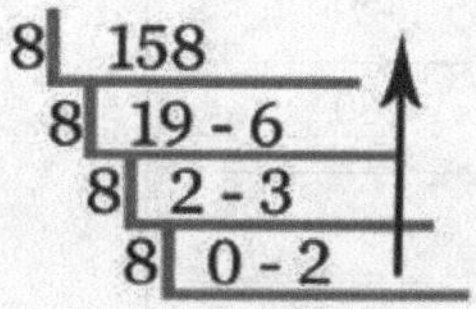

$(158)_{10} = (236)_8$

Therefore, $(158)_{10} = (236)_8$

- Rules for fraction number:
  - (i) Write the integer component you obtain after multiplying the fractional part by 8. Otherwise, type 0.
  - (ii) Once more, multiply the fraction portion of the result from the first step you obtained, write down the integer part, and repeat this process until the fraction part of the result is zero (minimum 4 to 5 times if it repeats).
  - (iii) Now, list your integers or zeros in the same order as you received them.

Example 1.19

Convert (0.45)10 to equivalent octal number.

Solution:

Multiplication	Result	Integer Portion	Fraction Portion
0.45 × 8	3.60	3	.60
0.60 × 8	4.80	4	.80
0.80 × 8	6.40	6	.40
0.40 × 8	3.20	3	.20
0.20 × 8	1.60	1	.60

Therefore, $(0.45)_{10} = (0.\ 34631..)_8$

Example 1.20

Convert $(123.45)_{10}$ to equivalent octal number.

Solution:

8 | 123
8 | 15 – 3
8 | 1 – 7
8 | 0 – 1

$(123)_{10} = (173)_8$

Multiplication	Result	Integer Portion	Fraction Portion
0.45 × 8	3.60	3	.60
0.60 × 8	4.80	4	.80
0.80 × 8	6.40	6	.40
0.40 × 8	3.20	3	.20
0.20 × 8	1.60	1	.60

Therefore, $(123.45)_{10} = (173.\ 34631..)_8$

1.4.6 Decimal-to-Hexadecimal Conversion

- Rules for converting integer:
  - (i) Calculate the reminder by dividing the decimal number by the hexadecimal number's base (16).
  - (ii) Divide the result by 16 once again, note down the reminder, and repeat this process until the result is 0.
  - (iii) The remainder, which is the hexadecimal counterpart of the provided decimal number, should now be written down in the opposite order.

Example 1.21

Convert $(450)_{10}$ to equivalent hexadecimal number

Solution:

16 | 450
16 | 28 - 2
16 | 1 - 12(C)
16 | 0 - 1

$(450)_{10} = (1C2)_{16}$

Therefore, $(450)_{10} = (1C2)_{16}$

- Rules for converting fraction:
  - (i) Write the integer component you obtain after multiplying the fractional part by 16. Otherwise, type 0.
  - (ii) Once more, multiply the fraction portion of the result from the first step you obtained, write down the integer part, and repeat this process until the fraction part of the result is zero (minimum 4 to 5 times if it repeats).
  - (iii) Now, list your integers or zeros in the same order as you received them.

1.4.7 Binary–Octal Conversion

- Rules for converting integer
  - (i) Take 3 bits from right to left and fill up by taking zeros if there is not available digit at the left to consist 3 bits.
  - (ii) Replace every 3 bit by their equivalent octal value.

Example 1.22

Convert $(0.78125)_{10}$ to equivalent hexadecimal number.

Solution:

Multiplication	Result	Integer Portion	Fraction Portion
0.78125 × 16	12.50	12(C)	.50
0.50 × 16	8.00	8	.00

Therefore, $(0.78125)_{10} = (C8)_{16}$

Example 1.23

Convert $(125.85)_{10}$ to equivalent hexadecimal number.

Solution:

16 | 125
16 | 7 - 13(D)
16 | 0 - 7

$(125)_{10} = (7D)_{16}$

Multiplication	Result	Integer Portion	Fraction Portion
0.85×16	13.60	12(D)	.60
0.60×16	9.60	9	.60
0.60×16	9.60	9	.60

Therefore, $(125.85)_{10} = (D99)_{16}$

Example 1.24

Convert $(1110101011)_2$ to equivalent octal number.

Solution:

001	110	101	011
1	6	5	3

Therefore, $(1110101011)_2 = (1653)_8$

- Rules for fraction number:
  - (i) Take every 3 bit from left to right and fill up by taking zeros if there is not available digit at the right to consist 3 bits.
  - (ii) Replace every 3 bit by their equivalent octal value.

Example 1.25

Convert $(0.10001000111)_2$ to equivalent octal number.

Solution:

.100	010	001	110
4	2	1	6

Therefore, $(0.10001000111)_2 = (0.4216)_8$

Example 1.26

Convert $(10010111.11001)_2$ to equivalent octal number.

Solution:

010	010	111	.110	010
2	2	7	.6	2

Therefore, $(10010111.11001)_2 = (227.62)_8$

1.4.8 Octal to Binary Conversion

- Rules for integer

(i) Every octal digit should be replaced with its three-bit binary counterpart.

(ii) In the same order as the octal number, write them one after the other.

Example 1.27

Convert $(723)_8$ to equivalent binary number.

Solution:

7 2 3

111 010 011

So, $(723)_8 = (111010011)_2$

Therefore, $(723)_8 = (111010011)_2$

- Rules for fraction number

(i) Every octal digit should be replaced with its three-bit binary counterpart.

(ii) In the same order as the octal number, write them one after the other.

Example 1.28

Convert $(0.043)_8$ to equivalent binary number.

Solution:

.0 4 3

.000 100 011

So, $(0.043)_8 = (0.000100011)_2$

Therefore, $(0.043)_8 = (0.000100011)_2$

1.4.9 Hexadecimal to Octal Conversion

- Rules for converting integer:

(i) Write the binary equivalent of each hexadecimal digit and its corresponding hexadecimal order next to it.

(ii) If there isn't a digit available to consist of 3 bits on the left, take every 3 bits from right to left and fill in with zeros.

(iii) Every 3 bit should be replaced with its octal equivalent.

- Rules for converting fraction:

(i) Write the binary equivalent of each hexadecimal digit and its corresponding hexadecimal order next to it.

(ii) If there isn't a digit available on the right to make up the third bit, fill in the gaps with zeros as you go from left to right.

(iii) Each 3 bit should be replaced with its corresponding hexadecimal value.

Example 1.29

Convert $(E5F)_{16}$ to equivalent octal number.

Solution:

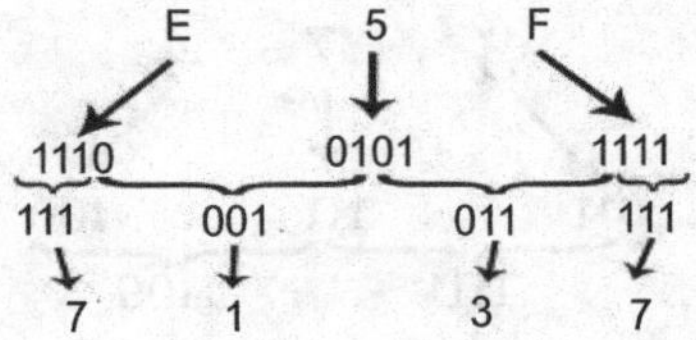

The Therefore, $(E5F)_{16} = (7137)_8$

Example 1.30

Convert $(0.7C9)_{16}$ to equivalent octal number.

Solution:

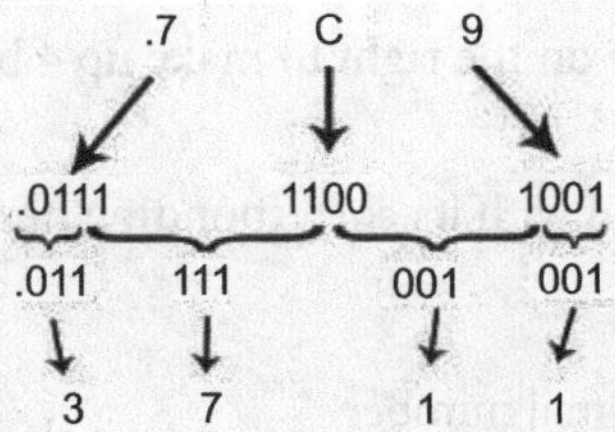

Therefore, $(0.7C9)_{16} = (0.3711)_8$

Example 1.31

Convert $(A4B.59E)_{16}$to equivalent octal number.

Solution:

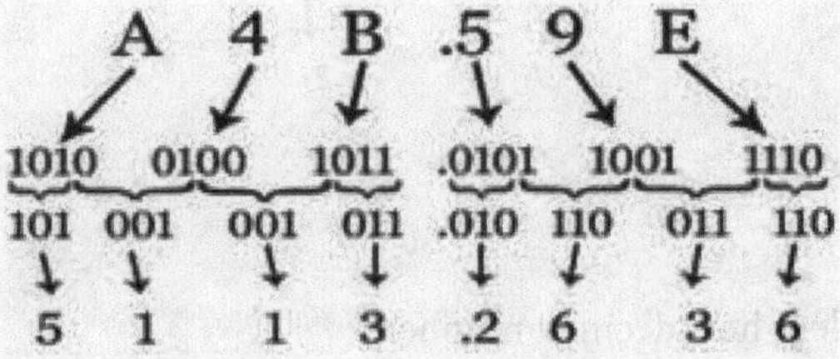

Therefore, $(A4B.59E)_{16} = (5113.2636)_8$

1.4.10 Octal to Hexadecimal Conversion

- Rules for converting integer:
  - (i) Replace each octal digit with its corresponding binary value, then write the results side by side in octal order.
  - (ii) If there isn't a digit available on the left to make up 4 bits, take each 4 bit from right to left and fill in with zeros.
  - (iii) Every 4 bits must be replaced with its corresponding hexadecimal value.

Example 1.32

Convert $(374)_8$ to equivalent hexadecimal number.

Solution:

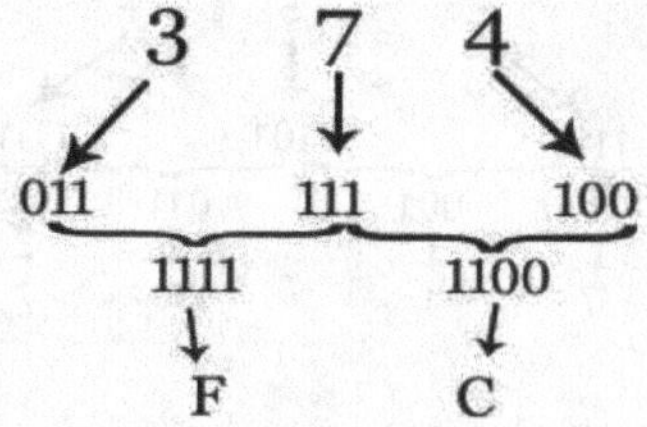

Therefore, $(374)_8 = (FC)_{16}$

- Rules for converting fraction:
  - (i) Replace each octal digit with its corresponding binary value, then write the results side by side in octal order.
  - (ii) If there isn't a digit available on the right to make up 4 bits, take each 4 bit from left to right and fill in with zeros.
  - (iii) Every 4 bits must be replaced with its corresponding hexadecimal value.

Example 1.33

Convert $(0.504)_8$ to equivalent hexadecimal number.

Solution:

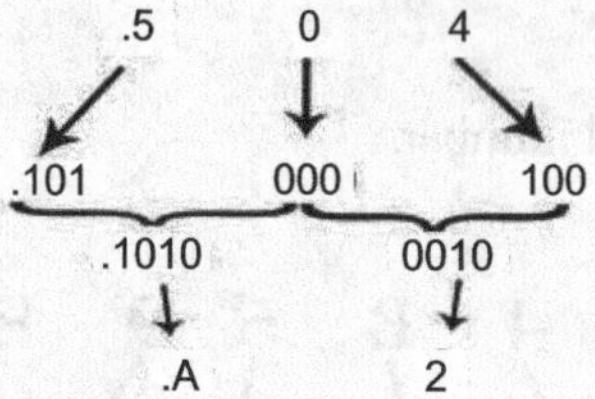

Therefore, $(0.504)_8 = (0.A2)_{16}$

Example 1.34

Convert $(343.405)_8$ to equivalent hexadecimal number.

Solution:

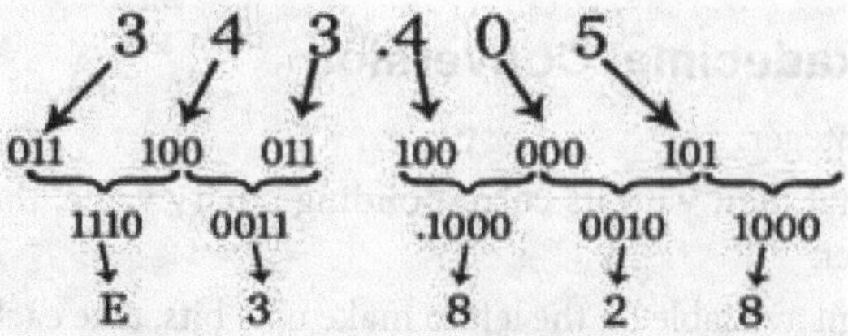

Therefore, $(343.405)_8 = (E3.828)_{16}$

1.4.11 Binary to Hexadecimal Conversion

- Rules for converting integer:

(i) Take each 4 bits from LSB towards MSB and fill in with zeros if there isn't a digit available on the left to make up 4 bits,

(ii) Every 4 bits must be replaced with its corresponding hexadecimal value.

- Rules for converting fraction:

(i) Take each 4 bits from MSB to LSB and fill in with zeros if there isn't a digit available on the right to make up 4 bits

(ii) Every 4 bits must be replaced with its corresponding hexadecimal value.

Example 1.35

Convert $(10111100111)_2$ to equivalent hexadecimal number.

Solution:

0101	1110	0111
5	E	7

Therefore, $(10111100111)_2 = (5E7)_{16}$

Example 1.36

Convert $(0.11001101001)_2$ to equivalent hexadecimal number.

Solution:

.1100	1101	1101
C	D	2

Therefore, $(0.11001101001)_2 = (0.CD2)_{16}$

Example 1.37

Convert $(10100110.10101)_2$ to equivalent hexadecimal number.

Solution:

1010	0110	.1010	1000
A	6	A	9

Therefore, $(10100110.10101)_2 = (A6.A9)_{16}$

1.4.12 Hexadecimal to Binary Conversion

- Rules for converting integer:

(i) Every hexadecimal digit should be replaced with its four-bit binary counterpart.

(ii) In the same order as the hexadecimal numbers, write them one after the other.

Example 1.38

Convert $(2B8)_{16}$ to equivalent binary number.

Solution:

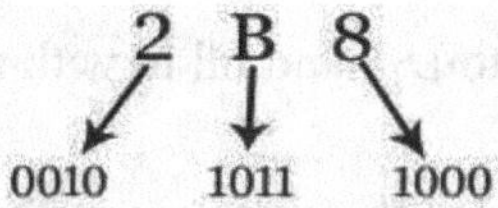

Therefore, $(2B8)_{16} = (001010111000)_2$

- Rules for converting fraction:
  - (i) Every hexadecimal digit should be replaced with its four-bit binary counterpart.
  - (ii) In the same order as the hexadecimal numbers, write them one after the other.

Example 1.39

Convert $(0.28C)_{16}$to equivalent binary number.

Solution:

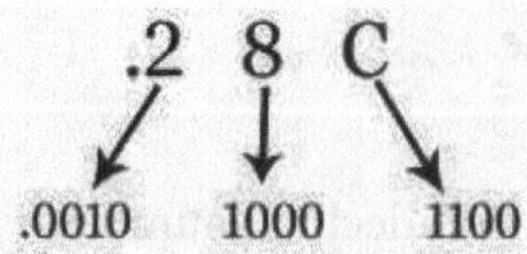

Therefore, $(0.28C)_{16} = (0.001010001100)_2$

Example 1.40

Convert $(CD4.28C)_{16}$ to equivalent binary number.

Solution:

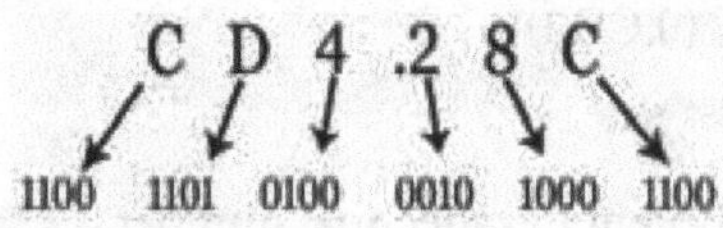

Therefore, $(CD4.28C)_{16} = (110011010100001010001100)_2$

1.5 BOOLEAN ALGEBRA

The category of algebra known as Boolean algebra uses variables to represent the true and false truth values which are often represented by the numbers 1 and 0, respectively. It is applied to simplify and examine digital gates or circuits. It is also known as logical algebra or binary algebra. It has played very important role in digital electronics advancement and is supported by all contemporary programming languages. **Boolean algebra is used to minimize a Boolean logic expression.** Another ways to minimize Boolean logic expression are**K-map and Quine McClusky tabular** method. Some important definitions related to Boolean algebra are given here-

Scan to know more More about Boolean algebra

Boolean or Logic variable – This is a logic variable that can take only values from set {0,1}.

Boolean or Logic function: Any functions which is made up of Boolean variables is known as Boolean function.

Example: F(A,B,C) = A+ BC

Here A,B, C are Boolean variable and F is Boolean function.

Boolean equation: An equation which consists of Boolean variable and/or {0,1} is known as Boolean equation.

Example: A + BC + ABC = 1

Truth Table- A truth table deconstructs a boolean expression by outlining all possible outcomes the function might achieve.

Logic Diagram: To represent any function in terms of basic gates or other gates is known as Logic Diagram.

A Boolean function can be represented either in Word statement/ Boolean equation/ Truth table or Logic diagram form.

1.5.1 Boolean Algebra Rules

(i) Only two values are possible for the utilised variable. Binary 0 is for LOW/FALSE and Binary 1 is for HIGH/TRUE.

(ii) An overbar is used to symbolise the complement of a variable. Consequently, the complement of variable B is denoted by B' or . Therefore, if B = 0 then B' = 1 and if B =1 then B'= 0.

(iii) The variable's OR operation is shown by a plus (+) symbol between them. (A + B + C), for instance, is used to express the OR of A, B, and C.

(iv) Writing a dot between two or more variables, such as A.B.C., represents a logical AND. Like ABC, the dot may occasionally be missed.

1.5.2 Laws of Boolean Algebra

Assume X, Y and Z are logic variables. So there are different governing Boolean algebraic laws as mentioned below-

1.5.2.1 Commutative Law

A commutative operation is any binary operation that meets the following equation. According to the commutative law, altering the order of the variables has no impact on the output of a logic circuit.

- X.Y = Y.X
- X + Y = Y + X

1.5.2.2 Associative Law

It claims that since their effects are the same, the sequence in which the logic operations are carried out is irrelevant.

- (X.Y).Z = X.(Y.Z)
- (X + Y) + Z = X + (Y + Z)

1.5.2.3 Distributive Law

It states that

- X. (Y + Z) = (X. Y) + (X. Z)
- X + (Y. Z) = (X + Y) . (X + Z)

1.5.2.4 AND Law

- X .0 = 0
- X. 1 = X
- X. X = X
- X.X'=0

1.5.2.5 OR Law

- X + 0 = X
- X + 1 = 1
- X + X = X
- X + X' = 1

1.5.2.6 Inversion Law

According to the inversion law, double inversion of a variable yields the original variable itself.

- X" = X

1.5.2.7 Complementation Law

(i) X .X' = 0
(ii) X +X' = 1

1.5.2.8 Idempotent Law

(a) X.X.X….X = X
(b) X +X +X +⋯+X = X

1.5.2.9 Absorption Law

(a) X +X.Y = X
(b) X.(X +Y) = X

1.5.2.10 Consensus Theorem

(i) X.Y +X'.Z +Y.Z = X.Y +X'.Z
(ii) (X +Y).(X' +Z).(Y +Z) = (X +Y).(X' +Z)

1.5.3 De-Morgan's Theorem

(i) (X.Y)' = X'+Y'
(ii) (X+Y)' = X'. Y'

1.5.1 Function Minimization using Boolean Algebra

Any Boolean function can be minimized using Boolean algebra. It can be easily understood by following examples.

Example 1.41

F = X + X'Y, minimize F using Boolean algebra.

Solution:

$$
\begin{aligned}
F &= X + X'Y \\
&= (X+X').(X+Y) \quad (\because \text{ by distributive law}) \\
&= 1.(X+Y) \quad (\because \text{ by complement law } A + A' = 1) \\
&= X + Y
\end{aligned}
$$

Example 1.42

F = X.(X'+Y), minimize F using Boolean algebra.

Solution:

$$
\begin{aligned}
F &= X.(X'+Y) \\
&= X.X' + X.Y \quad (\because \text{ by distributive law}) \\
&= 0+X.Y \quad (\because \text{ by complement law } A.A' = 0) \\
&= X.Y \text{ is the solution}
\end{aligned}
$$

1.6 KARNAUGH MAPS AND THEIR USE FOR SIMPLIFICATION OF BOOLEAN EXPRESSIONS

Karnaugh Map or K-map is the other Boolean Logic minimisation method. This is most widely used method. This method brings the minimized expression in the solution. In the minimised expression number of literal (number of variables appeared in final expression) are minimised. K-map can be prepared for 2 or more variables as shown in the Figure 1.2 and Figure 1.3 below. For further reading please scan the **QR code** given in next page.

Scan to know more about K-map

K-maps can be prepared with minterms or maxterms.

Minterm- it is a product term which includes all variables in prime or complement form. Minterm is represented by mi, where 'i' is the decimal equivalent of that binary number.

In Minterms, '1' is treated as 'prime' while '0' is treated as 'complement' for variables as well as functions. We need to make pairing of 1's and the final result is sum-of product (SOP).

Maxterm- It is a sum term which includes all variables in prime or complement form. Maxterm is represented by Mi, where 'i' is the decimal equivalent of that binary number.

In Maxterms, '0' is treated as 'prime' while '1' is treated as 'complement' for variables as well as functions. We need to make pairing of 0's and the final result is product-of-sums (POS).

Scan to know more about Implicants

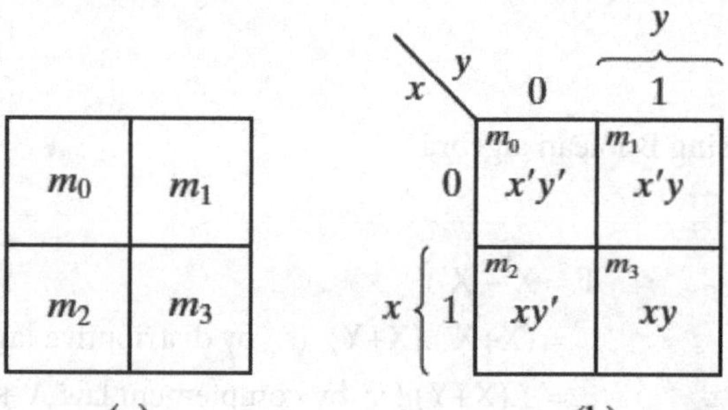

Fig. 1.2: Two-variable Karnaugh map

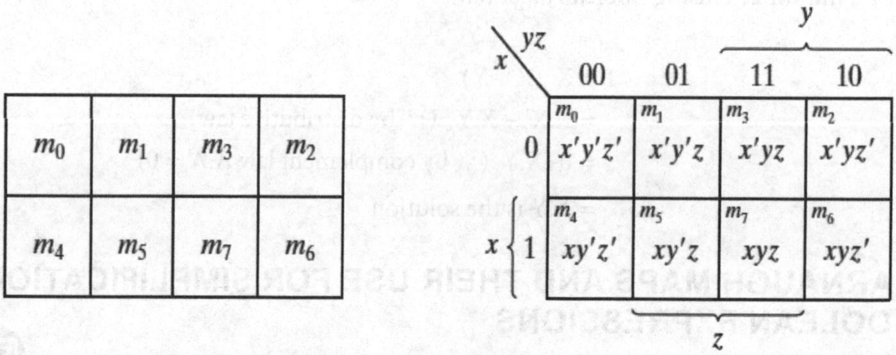

Fig. 1.3: Three-variable Karnaugh map

- The procedures to achieve a simplified SOP solution using K-map are as follows.
  - (i) The expression should be expressed in its standard form.
  - (ii) Fill the K-map with data -For each product-term, enter "one" in the K-map cell and "0" in the adjacent cells. Fill 'X' for DON'T care terms.
  - (iii) Create Groups by combining adjacent '1' and if needed use 'X' for grouping.
  - (iv) Write the expression using only CONSTANT variables in that group.
  - (v) See if any '1' is not covered write minterm for this in final expression and remove redundant grouped product terms from the final expression.
- Rules for grouping

A group can be prepared for 2*i* Minterms or Maxterms

Example: Simplify Y = AB + A'B + A'B'

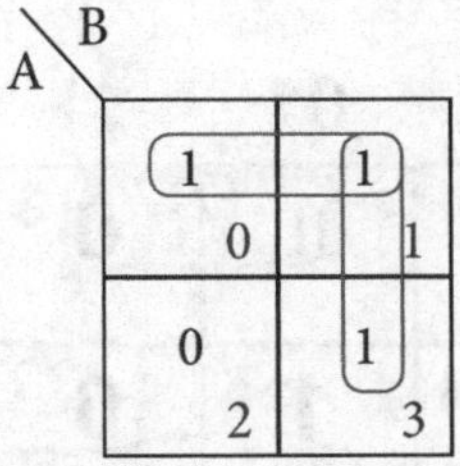

Y = B + A'

Example: Simplify Y = ABC + AB'C + A'B'C' + A'BC' + AB'C' + ABC'

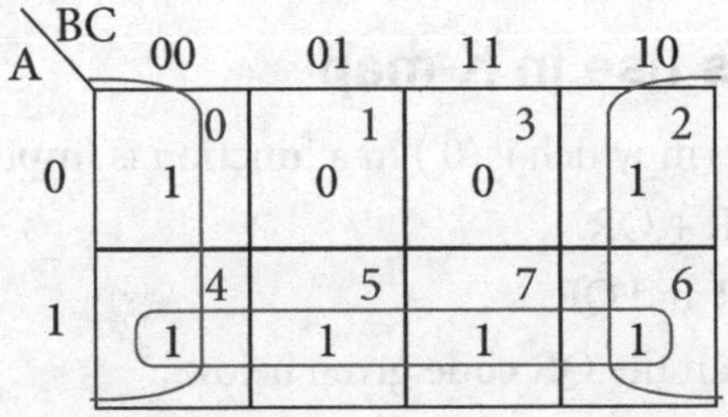

Y = C' + A

Example: Simplify the logic expression

Y = ABCD + A'BCD + A'BCD' + A'B'C'D' + A'B'CD' + AB'C'D' + AB'CD' + ABCD'

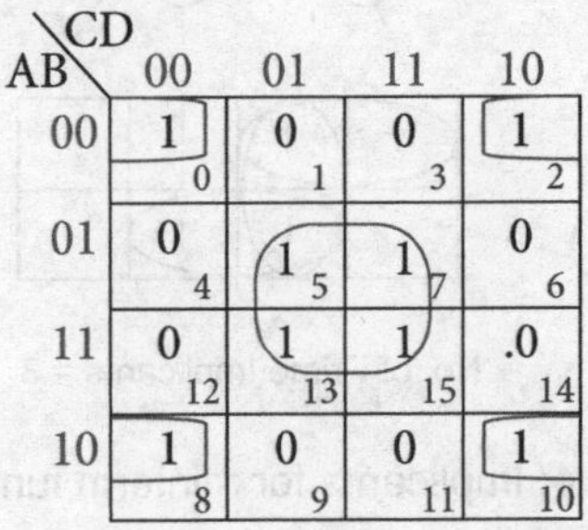

Y = B'D' + BD

- Maxterm Solution of K Map
  - (i) For each sum-term of the expression, "0" rather than "1" are to be entered into the K-map cells.
  - (ii) It is recommended to continue grouping "0" and DON'T cares (X) if needed rather than "ones."
  - (iii) Each group's Boolean expression must be written as **sum-term** rather than product-term.
  - (iv) Any redundant group has to be removed and any uncovered Maxterm has to be included in the final expression.

Example: Y = (A' + B' + C') + (A + B + C') + (A + B' + C') + (A' + B' + C)

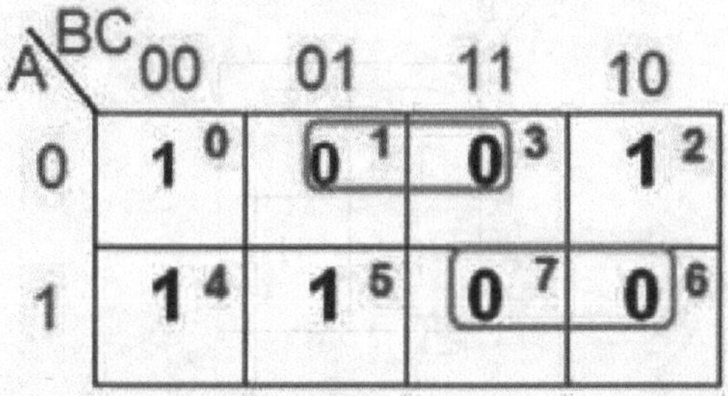

Y = (A' + B') · (A + C')

1.6.1 Implicants and its use in K-map

Every minterm which is '1'(maxterm which is '0') in a function is **implicant**.

E.g., function, F = PQ + PQR + QR.

Here Implicants are PQ, PQR and QR.

For further reading please scan the **QR code** given below.

Scan to know more about Implicants

Prime Implicants –

All possible pairing of implicants (minterms) in K-Map are called **prime implicants (PI)**

Example: There are three possible prime implicants in the below figure.

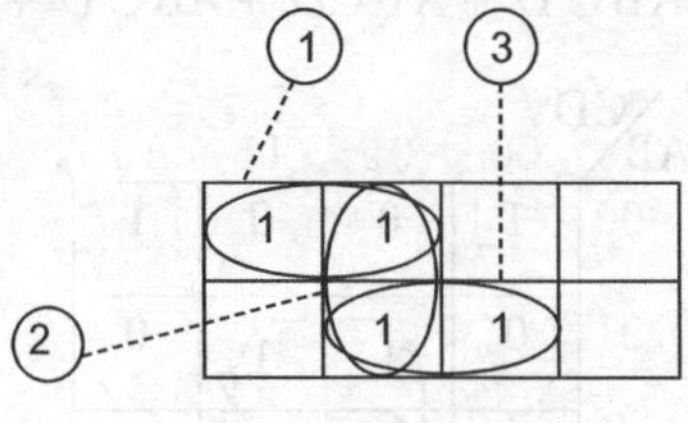

No. of Prime Implicants = 3

Fig. 1.4: Implicants for minterm function

Essential Prime Implicants: These groups include at least one minterm(or maxterm in case of POS) that is not covered by any other prime implicant.

Example: Here in fig. 1.4., Prime implicants 1 and 3 are Essential Prime implicants.

Redundant Prime Implicants: If minterms are included by some essential prime implicant and still there are more Prime Implicants. These extra prime implicantsare **Redundant prime implicants(RPI).**

Example: Here in fig. 1.4., Prime implicants 2 is Redundant Prime implicant, this is not needed in the final solution as the 1's covered by it are already covered by Essential Prime implicants 1 and 3.

Selective Prime Implicants Selective prime implicants(SPI) **are Redundant prime implicants but some of these are included the final solution.**

Example: Here in the below example prime implicants 1 and 2 are Selective Prime implicants but these are neither essential nor redundant.

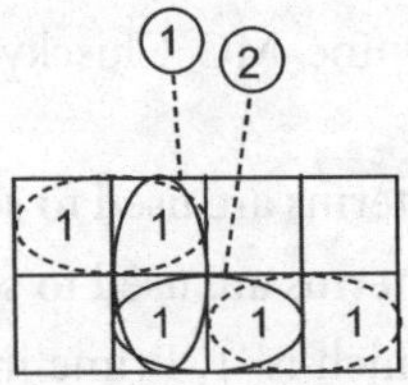

No. of Selective Prime Implicants = 2

Example: Given F = Σ(1, 5, 6, 7, 11, 12, 13, 15), find number of implicant, PI, EPI, RPI and SPI.

Solution:

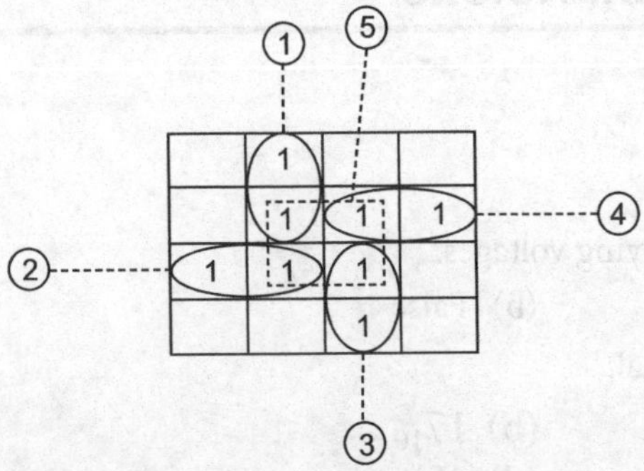

No. of Implicants = 2

PI = (1, 2, 3, 4, 5)

EPI = (1, 2, 3, 4)

RPI = (5)

F = ① + ② + ③ + ④

No. of Imlicatns = 8

No. of Prime Implicatns (PI) = 5

No. of Essential Prime Implicants (EPI) = 4

No. of Redundant Prime Implicants (RPI) = 1

No. of Selectiv e Prime Implicants (SPI) = 0

UNIT SUMMARY

- Analog signals are continuous in time and amplitude domain while digital signals are continuous in time domain but only at '0' and '1' level in amplitude domain.
- Binary number system is used by processors while decimal number system is easy to access by human beings.
- Binary to octal conversion can be done by grouping three bits and writing equivalent Octal number for that, grouping has to start from decimal.
- Binary to Hexadecimal conversion can be done by grouping four bits and writing equivalent Hexadecimal number for that, grouping has to start from decimal.
- Any number to decimal can be obtained by multiplying positional weights and face values.
- Boolean algebra deals with Boolean functions.

- Boolean algebra, K-map and Quine MC Cluscky are methods to minimise any Boolean function.
- K-map can give SOP form is minterms are used to solve. Grouping of '1'.
- K-map can give POS form if maxterms are used to solve. Grouping of '0'.
- Implicant means any minterm which is '1', Prime implicant is a group of implicants which is prepared by k-map rules
- Essential Prime Implicants are always part of final solution along with some selective prime implicants

EXERCISES

Examples are given along with topics.

Multiple Choice Questions

1. Digital electronics use continuously varying voltages.
(a) True (b) False

2. Convert hexadecimal value 17 to decimal.
(a) 22_{10} (b) 17_{10}
(c) 11_{10} (d) 20_{10}

3. A 2's complement of any binary number'scan be obtained adding '1' to it's 1's complement.
(a) True (b) False

4. Find binary equivalent of $(C1)_{16}$.
(a) 11000001 (b) 1000111
(c) 11100010 (d) 11100001

5. Find Octal number for $(010111100)_2$
(a) $(782)_8$ (b) $(501)_8$
(c) $(234)_8$ (d) $(274)_8$

6. A bit with the 1 sign represents a positive binary number.
(a) True (b) False

7. What is binary equivalent for 45_8?
(a) 011010 (b) 100101
(c) 100101 (d) 010100

8. Which is the digit set for binary number system?
(a) {H, L} (b) {T, F}
(c) {1, 2} (d) {0, 1}

9. The range of an analogue signal is 0 to 5 V. How many different analogue configurations are there overall in this space?
(a) 15 (b) 500
(c) 50 (d) infinite

10. What is 2's complement of binary number 11100101?

(a) 00100110 (b) 00011011
(c) 11011000 (d) 10010010

11. The repeated division-by-2 technique can be used to translate a decimal fraction into binary.

(a) True (b) False

12. Hexadecimal 44 IS equivalent to $( )_2$

(a) 01000100 (b) 11101010
(c) 11010001 (d) 10101001

13. The 2's complement of 1000 is ________.

(a) 1110 (b) 1100
(c) 0100 (d) 1000

14. When converting from binary to hexadecimal, zeros may be added to the left of the MSB to make even groups of 4 bits.

(a) True (b) False

15. According to the commutative property of Boolean algebra A + B equals A.B.

(a) Yes (b) No

16. The expression $\overline{ABC}$ is equal ……..using DeMorgan's theorem.

(a) $\overline{A} + \overline{B} + \overline{C}$ (b) $\overline{A + B + C}$
(c) $A + \overline{B} + C\overline{C}$ (d) $A(B + C)$

17. Find values of P, Q, R, and S such that P'QR'S =1.

(a) P = 0, Q = 1, R = 0, S = 1 (b) P = 0, Q = 0, R = 0, S = 1
(c) P = 1, Q = 1, R = 1, S = 1 (d) P = 0, Q = 0, R = 1, S = 0

18. PR + PQR = PR

(a) True (b) False

19. The expression $\overline{\overline{(X + Y)} + \overline{Z}}$ can be simplified as ______.

(a) $(X + Y)Z$ (b) $(\overline{X} + \overline{Y})Z$
(c) $(X + Y)\overline{Z}$ (d) $(\overline{X} + \overline{Y})\overline{Z}$

20. The expression $A(B + \overline{C} + D)$ can be represented as______.

(a) AB + AC + AD (b) ABCD
(c) A + B + C + D (d) $AB + A\overline{C} + AD$

21. The expression $\overline{(W + X + Y)\, Z}$ can be represented as ________.

(a) $\overline{W}\,\overline{X}\,\overline{Y}\,\overline{Z}$ (b) $\overline{WXY}\,Z$
(c) $WXY\overline{Z}$ (d) $\overline{W}\,\overline{X}\,\overline{Y} + \overline{Z}$

22. The distributive law of Boolean algebra is :

(a) (P .Q) . R = P . (Q . R) (b) P+(Q .R) = (P+Q).(P+R)
(c) P + (Q + R) = (P+Q).(P+R) (d) P+(QR) = (P+Q) . R

23. The expression $\overline{A + B + C + D}$ can be written as ______.
(a) $\overline{A}\,\overline{B}\,\overline{C}\,\overline{D}$
(b) $\overline{A} + \overline{B} + \overline{C} + \overline{D}$
(c) $A + \overline{B} + \overline{C} + D$
(d) $\overline{A} + B + C + \overline{D}$

24. In Boolean algebra, $\overline{\overline{A}} = A$.
(a) Yes
(b) No

25. Which of the following illustrations best captures the commutative law of multiplication?
(a) $P + Q = Q \cdot P$
(b) $P \cdot Q = Q + P$
(c) $P \cdot Q = Q \cdot P$
(d) $P \cdot Q = P \times Q$

ANSWERS															
1.	(b)	2.	(a)	3.	(a)	4.	(a)	5.	(d)	6.	(b)	7.	(b)	8.	(d)
9.	(d)	10.	(b)	11.	(b)	12.	(a)	13.	(d)	14.	(a)	15.	(b)	16.	(a)
17.	(a)	18.	(a)	19.	(a)	20.	(d)	21.	(d)	22.	(b)	23.	(a)	24.	(a)
25.	(c)														

Short and Long Answer Type Questions

1. Find binary for the following
(a) $(E3FA)_{16}$
(b) $(202)_6$
(c) $(145)_{10}$
(d) $(107)_8$

2. Simplify F = ((A + BC')' + D) 'using Demorgan'slaw.

3. Find the 2's complement and 1's complement of binary number 110110101.

4. Simplify -
(a) A +AB
(b) A + (B.A')
(c) A + AB + AC
(d) A + B + AB

Numerical Problems

1. Convert the following:
(i) $(101010)_2 = ()_{10}$
(ii) $(11100)2 = ()_{10}$
(iii) $(10001)_2 = ()_{10}$
(iv) $(111)_2 = ()_{10}$
(v) $(10512.145)8 = ()_{10}$
(vi) $(57.4)_{16} = ()_{10}$
(vii) $(BAD)_{16} = ()_{10}$
(viii) $(EBF1.A2)_{16} = ()_{10}$
(ix) $(47)_{10} = ()_2$
(x) $(1234.567)_{10} = ()_2$
(xi) $(394)_{10} = ()_8$
(xii) $(832.91)_{10} = ()_8$

(xiii) $(542)_{10} = ()_{16}$
(xiv) $(176.32)_{10} = ()_{16}$

2. Simplify the following:
   (i) (P' +Q) (P'+ R') (P +R)
   (ii) P'QR' + P Q' R' + PQ R'
3. Minimize using Boolean rules.
   (i) F = P.Q.R + P'.Q + P.Q.R'
   (ii) F = P'.Q.R' + P.Q'.R' + P'.Q'.R' + P'.Q'.R'
   (iii) P.Q + P'.R + Q.R = P.Q + P'.R
   (iv) (P + Q).(P' + R).(Q + R) = (P + Q).(P' + R)
4. Minimize using k-map
   F = Q' R' + Q 'S' + PQ' + PS + PR + RS'
5. Minimize using k-map
   F = P.R'.S + P'.Q.R'.S + P'.Q'.S + P.Q'.R.S
6. Apply DeMorgan's theorems:
   (a) (X + Y + Z) P
   (b) XYZ + PQR
   (c) XY + PQ + MN
7. Simplify the following:
   1- X.Y' + X.(Y + Z) + Y.(Y + Z).
   2- (X.Y'.(Z + Y.Z) + X'.Y').Z
   3- X'.Y.Z + X.Y'.Z' + X'.Y'.Z' + X.Y'.Z + X.Y.Z

PRACTICAL

Laboratory Do's and Don'ts

1. Arrive at the laboratory experiment completely prepared.
2. Before connecting to the equipment, make sure that the power supply is appropriate.
3. Based on the quantity to be measured, choose the right range of measuring devices.
4. Connect the wires without attaching the supply's leads. Before switching-on the power supply to the circuit, re-check the connections and show it to the teacher /instructor.
5. Only energise the circuit with the instructor's or teacher's approval.
6. After the experiment, detach the wires and leads and reposition them where they belong.
7. Give the lab workers all the equipment.
8. Immediately turn off the power supply in the event of a shock.
9. Adhere strictly to the instructions provided with the relevant experiments.

10. Avoid touching body earth and the main power supply leads with bare hands.

11. Avoid using mobile devices in the lab.
    - Experiments are started from Chapter 2 onwards when students will have learned about basic gates.
    - Experiments can be done in the lab using digital ICs.
    - Experiments can be done on EDA Playground website using VERILOG language (Login is required using email ID).
    - Experiments can be performed in virtual labs as well. Here QR code for Virtual labs from IIT Bombay and IIT Kharagpur are provided respectively.

Virtual - Lab IIT Bombay

Virtual - Lab IIT Kharagpur

KNOW MORE

Basic building block for digital electronics was BJT earlier but now a days MOSFET is the key component for all digital logics. BJT based logics were high power consuming but MOSFET based CMOS logics have very low power in standby modes although dynamic power is consumed.

REFERENCES AND SUGGESTED READINGS

- Digital Design by M. Moris Mano and Michael D Ciletti, 5th edition, Pearson
- Digital Electronic Circuits - Course (nptel.ac.in)

Dynamic QR Code for Further Reading

- QR codes embedded in the chapter

2 Logic Gates

UNIT SPECIFICS

Through this unit we have discussed the following aspects:

- Positive and Negative Logic;
- Truth Table;
- Logic Gates: Symbolic representation and truth table Implementation of Boolean expressions and Logic Functions using Gates;
- Universal Gates Simplification of expressions

The practical applications of the topics are discussed for generating further curiosity and creativity as well as improving problem solving capacity.

Besides giving a large number of multiple choice questions as well as questions of short and long answer types marked in two categories following lower and higher order of Bloom's taxonomy, assignments through a number of numerical problems, a list of references and suggested readings are given in the unit so that one can go through them for practice. It is important to note that for getting more information on various topics of interest some QR codes have been provided in different sections which can be scanned for relevant supportive knowledge.

After the related practical, based on the content, there is a "Know More" section. This section has been carefully designed so that the supplementary information provided in this part becomes beneficial for the users of the book. This section mainly highlights the initial activity, examples of some interesting facts, analogy, history of the development of the subject focusing the salient observations and finding, timelines starting from the development of the concerned topics up to the recent time, applications of the subject matter for our day-to-day real life or/and industrial applications on variety of aspects, case study related to environmental, sustainability, social and ethical issues whichever applicable, and finally inquisitiveness and curiosity topics of the unit.

RATIONALE

This fundamental unit on logic gates helps students to get a primary idea about the Boolean logic expressions, also known as logic expressions which are implemented using logic gates and termed as digitalelectronic circuits. The most fundamental component of combinational logic is the logic gate. There are three basic or fundamental logic gates NOT gate, AND gate, &OR gate. There are other derived gates like XOR gate, XNOR gate, NAND gate & NOR gate, which are derived using basic gates. This chapter discusses the fundamental functionalities of all digital logic gates and other related devices, like drivers, buffers etc. All combinational circuits are mostly handled with the use of appropriate Boolean expressions, logic diagrams andtruth tables. Solved examples are used to effectively explain the chapter.

PRE-REQUISITES

Basic concept Boolean logic and K-map.

UNIT OUTCOMES

List of outcomes of this unit is as follows:

U2-O1: Describe positive and negative logic of the gates

U2-O2: Describe truth table and its implementation

U2-O3: Describe types of logic gates (AND, OR, NOT, NOR, XNOR)

U2-O4: Symbolic representation using logic gates

U2-O5: Explain other gates in terms of Universal gates

Unit-2 Outcomes	EXPECTED MAPPING WITH COURSE OUTCOMES (1 - Weak Correlation; 2 - Medium Correlation; 3 - Strong Correlation)					
	CO-1	CO-2	CO-3	CO-4	CO-5	CO-6
U2-O1	1	3	3	-	-	1
U2-O2	2	3	3	2	-	-
U2-O3	3	3	3	2	-	1
U2-O4	2	3	3	2	-	1
U2-O5	3	3	3	2	-	1

"Success is the sum of small efforts, repeated day in and day out."

Unknown

2.1 INTRODUCTION

Typically, when referring to a decision-making process, the word "logic" is employed. So a logic gate is a circuit that can decide whether to output a Yes or No depending on the inputs. Digital logic levels (ones and zeros) are combined in specified ways using gates, which are circuits. A logic gate serves as the fundamental building component of every digital circuit. The outcome is expressed in terms of the inputs using a system called Boolean Algebra and the related tabular representation known as a Truth Table. A gate is a digital circuit that has a single output signal but one or more input signals. A gate is an electrical switching circuit that, given specific logical conditions, permits the application of an input signal to be passed. The fundamental gates are NOT, AND, and OR gates. Universal logic gates NAND and NOR are created from an AND gate. The NOT gate is followed by the AND gate in the NAND gate. A NOR gate is a NOT gate followed by an OR gate, similarly. Exclusive NOR gate (XNOR) and Exclusive OR (XOR) gate are two additional fundamental logic gates (XNOR).

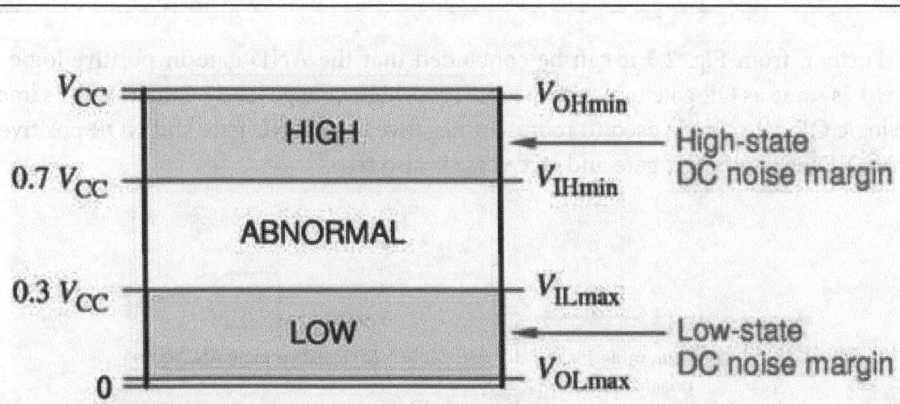

Fig. 2.1: Logic levels and noise margins for CMOS devices

Scan to know more about Gates

There are different types of logic families like CMOS- Complementary Metal Oxide Semiconductor (lowest power), ECL- emitter coupled logic (fastest one), TTL-transistor transistor logic etc. CMOS based digital circuits have both NMOS and PMOS in equal numbers and consume least power among all other logic families, therefore, CMOS based logics are more popular. As shown in Fig. 2.1, the logic level '0' is considered from V_{OL} to V_{IL} and likewise logic '1' is considered from input voltage level V_{IH} to V_{OH}. Therefore, actually there is a voltage range for inputs '0' or LOW level and '1' or HIGH level. This range is known as noise margin. There are two types of Noise Margins (NM)- low NM – $NM_L = V_{IL} - V_{OL}$ and High NM – $NM_H = V_{OH}$ to V_{IH}. If the input voltage is anywhere between V_{IL} to V_{IH} this is neither considered as '0' nor as '1' and this range VIL to VIH is known as FORBIDDEN or abnormal range. Forbidden range is not used in digital logic circuits.

Fig. 2.2: (a) Positive logic (b)Negative logic

2.1 POSITIVE AND NEGATIVE LOGIC

As shown in Fig. 2.2, binary variables can exist either logic '0' or '1' state. In digital systems, like processors, these logic states are represented by two distinct current levels or voltage levels as shown in Fig. 2.1.If high-(voltage or current) level H is used to represent logic '1' and low level L is used to represent logic '0', this is defined as positive logic system. While if the LOW--(voltage or current) level L is used to represent logic '1' and high level H is used to represent logic '0', this is defined as negativelogic system. Fig. 2.2 represents both positive and negative logic systems.

Fig. 2.3, represents positive and negative logic gates truth table and symbolic diagrams. Here in negative logic OR gate small triangles at inputs and outputs are there to represent that it is a negative logic gate. Further, from Fig. 2.3 it can be concluded that the AND gate in positive logic (1=> high voltage level) is same as OR gate in negative logic (0=> high voltage level). Likewise the same hardware of positive logic OR gate can be used to represent negative logic AND gate and NOR positive logic gate is same as NAND negative logic gate and vice versa is also true.

x	y	z
0	0	0
0	1	0
1	0	0
1	1	1

(a) Truth table for positive logic

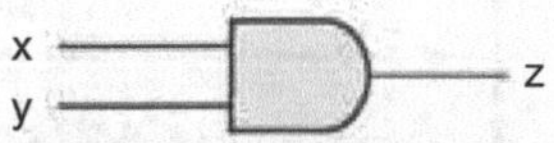

(b) Positive logic AND gate

x	y	z
1	1	1
1	0	1
0	1	1
0	0	0

(a) Truth table for negative logic

(b) Negative logic OR gate

Fig. 2.3: Truth table and symbolic diagrams for positive logic AND gate (a),(b) and negative logic OR gate (c), (d)

A	B	A and B	A or B
T	T	T	T
T	F	F	T
F	T	F	T
F	F	F	F

T=true F=false

Truth Table			
C	B	A	Q
0	0	0	0
0	0	1	0
0	1	0	0
0	1	1	0
1	0	0	0
1	0	1	0
1	1	0	0
1	1	1	1

Fig. 2.4: (a) Two-input logic system AND, OR truth table (b) Three-input AND logic system truth table

2.2 TRUTH TABLE

A truth table is one way to represent any Boolean logic circuit, other ways to represent are Boolean expressions and logic diagrams. Truth table for n-input logic circuit will have 2n input entries and corresponding outputs for each entry.Boolean expression is a logic equation for outputs in terms of input logic variables.Thus, there are just two (21) possible inputs, "0" and "1," when there is only one binary input variable. There are four (22) possible input combinations, namely 00, 01, 10 and 11, if there are only two inputs. The 2 input system's truth table is shown in Fig. 2.4. (a) for basic AND gate & OR gate.A three-input logic circuit's truth table is shown in Fig. 2.4 (b), contains 8 (= 23) rows, here output

Q=1 only iff all three inputs are '1'. Thus this is the truth table for 3-input AND gate. To find out the final expression from truth table we can use K-map to solve it either using minterms (SOP form) or maxterms (POS form).

2.3 LOGIC GATES

Every digital system like processor/ computer is having logic gate as basic building block. Although tosimplify any Boolean logic expression, Boolean algebra rules are applied, the actual implementation of these logic expressions in any digital logic system makes use of smaller logic circuits are called logic gates. There are total 8 logic gates namely- AND, OR, NOT, XOR, XNOR, NAND, NOR and Buffer. NOT gate, OR gate, AND gate, -are three basic logic gates, because any Boolean logic expression can be implemented using these three gates only.NAND gate & NOR gate -are two universal gates, as any Boolean expression can be represented using NAND gates alone or NOR gates alone. XOR & XNOR gates are derived gates while buffer is back to back two NOT gates are connected. Buffer is mainly used for providing delay. XOR & XNOR are mainly used to find even or odd number of 1's in the input bit set. All of these gates are mentioned below.

Please note, unless until not specified assume that all logics gates are with POSITIVE logic. Also note that all the logic gates are represented always with capital letters. Fig. 2.6 represents all gates, Boolean expressions, symbols and truth tables.

2.3.1 AND Gate

An AND gate hasone output and two/more inputs. In AND gate, the output is HIGH if and only if all the applied input bits are HIGHelse the output is LOW. A positive logic AND gate's output is "1" if and only if all its inputs are "1" else the out is '0'. Fig. 2.3(a), (b) show the truth table and the logic symbol for a two-input AND gate, respectively, also shown in Fig. 2.6. z = x.yrepresents the ANDing operation x and y. Here, z is the output result, and x and y are the input logic variables. Likewise, Fig. 2.4(b) represents AND gate's truth table for three inputs. The output Q = ABC according to the given truth table. To obtain this output once can use K-map to solve it.

Y = AB (for2 inputs);

Y = ABC (for 3 inputs);

Y = A1. A2..................An (for ninputs-generalised);

Where A,B,C or Ai's are the inputs.

Scan to know more about Gates

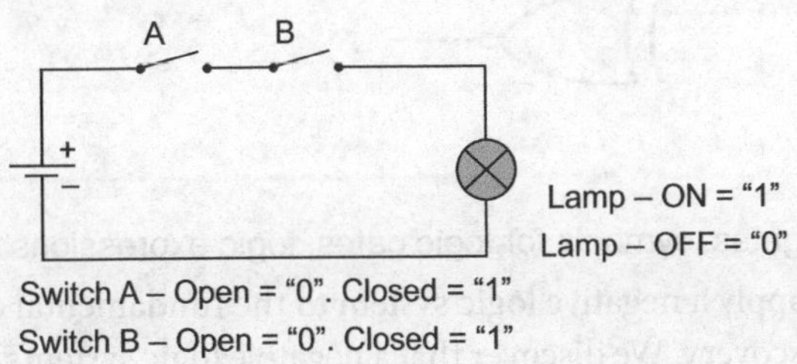

Fig. 2.5: Electrical analogy of two-input AND gate

Name	Graphic symbol	Algebraic function	Truth table
AND	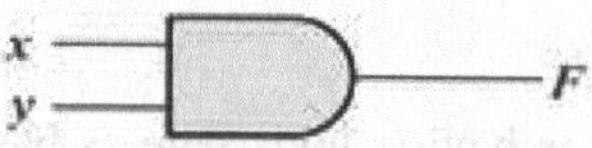	$F = x \cdot y$	x y F 0 0 0 0 1 0 1 0 0 1 1 1
OR		$F = x + y$	x y F 0 0 0 0 1 1 1 0 1 1 1 1
Inverter	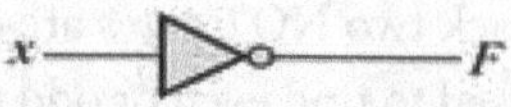	$F = x'$	x F 0 1 1 0
Buffer	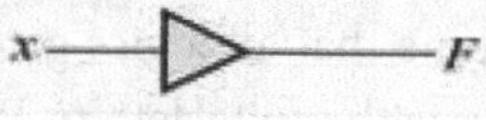	$F = x$	x F 0 0 1 1
NAND	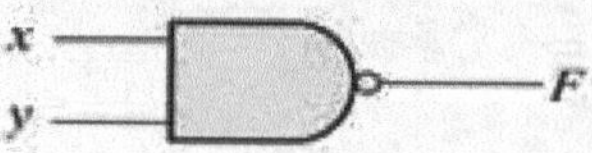	$F = (xy)'$	x y F 0 0 1 0 1 1 1 0 1 1 1 0
NOR	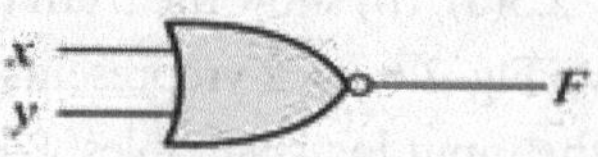	$F = (x + y)'$	x y F 0 0 1 0 1 0 1 0 0 1 1 0
Exclusive-OR (XOR)	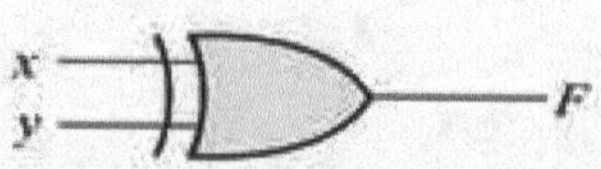	$F = xy' + x'y$ $= x \oplus y$	x y F 0 0 0 0 1 1 1 0 1 1 1 0
Exclusive-NOR or equivalence	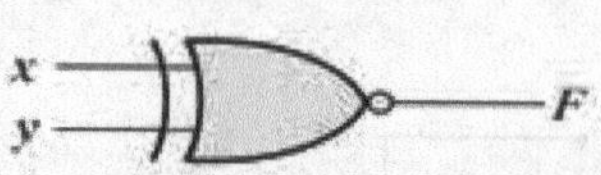	$F = xy + x'y'$ $= (x \oplus y)'$	x y F 0 0 1 0 1 0 1 0 0 1 1 1

Fig. 2.6: Logic gates, symbols for logic gates, logic expressions and truth tables

KNOW MORE-If we apply a negative logic system to the fundamental definitions of OR and AND gates, we get an amazing discovery. We discover that a negative logic system's AND gate is a positive logic system's OR gate. Additionally, a negative OR is a positive AND.

Electrical Analogue:

Electrical analogue is given in Fig. 2.5. When both switches are closed (=> inputs A='1' and B='1'), the bulb receives current, and the output is '1' (=> bulb is glowing). Ff any switch is left unlocked (=> either A='0' or B='0'), the circuit becomes an open circuit and the blub is OFF that means output is '0'.

2.3.2 OR Gate

Logical expression for two input OR gate is represented as Y = A + B, which read as Y = A OR B where A, B are the inputs and Y is the output. OR gate can have one output and two/more inputs. In OR gate output be LOW if all of an inputs are LOWelse the output is HIGH for all other input configurations. So for positive logic OR gate if all inputs are at logic '0'the output logic is '0' else the output is a logic "1" for all other potential input combinations. Fig. 2.6 represents a two-input OR gatewith symbol, logical expression and truth table. For two input OR gate logic statement is F = x + y. Likewise three input OR gate is presented in Fig.2.7, where Q = A + B + C.

Symbol	Truth Table			
	C	B	A	Q
	0	0	0	0
	0	0	1	1
	0	1	0	1
	0	1	1	1
	1	0	0	1
	1	0	1	1
	1	1	0	1
	1	1	1	1
Boolean Expression Q = A+B+C	Read as A OR B OR C gives Q			

Fig. 2.7: Three input OR gate symbols and truth tables

Electrical Analogue:

When both switches are open (=> AB=00), current does not flow through the lightbulb, producing '0' at the output. If any one of the switch is closed or both switches are closed i.e. AB = 01, 10 or 11, it allows electricity to flow via the short circuit as shown in Fig. 2.8. Thus making the bulb ON (=> output '1'). Therefore OR gate's truth table is verified.

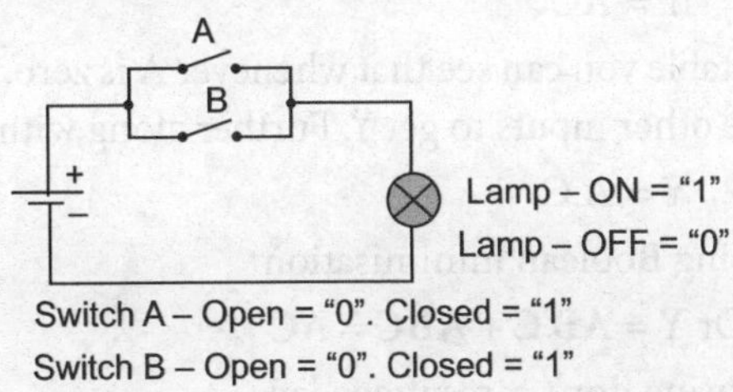

Fig. 2.8: Electrical analogue of OR gate.

2.3.3 Inverter or Not Gate

A logic gate with a single input and output known as a NOT gate always outputs the input's complement. Consequently, a LOW input results in a HIGH output, and vice versa. NOT gate is also known as inverter, since it INVERTs (complements) the input bit. A positive logic system is one in which a logic '0' at the input results in a logic '1' at the output, and the other way around. As a "inverting circuit" or "complementing circuit," it also goes by those names. In Figure 2.9, the truth table and circuit symbol, symbols are presented for NOT gate.

The symbol for the NOT operation on a logic variable X is $\overline{X}$ (some other representations are X' or !X or ~X). To put it another way, if X is the input to a NOT circuit, then Y, the output, is determined by $Y = \overline{X}$., which translates to Y equals NOT X. As a result, if X = 1 then Y = 0 and vice versa.

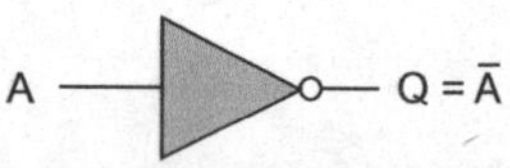

Inverter on NOT Gate

Switch	Output
1	0
0	1
Boolen Expression	not-A or $\overline{A}$

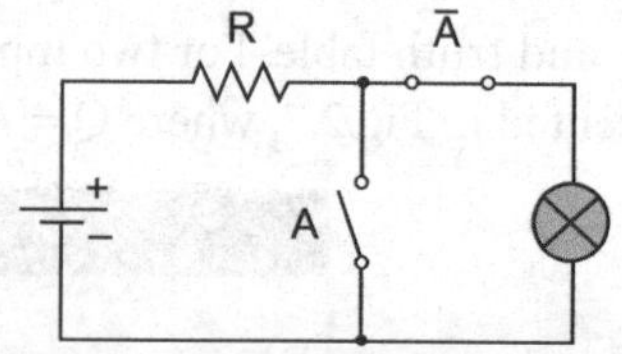

Switch A – Open = "0". Lamp – ON = "1"
Switch A – Closed = "1". Lamp – OFF = "0"

Fig. 2.9: Symbol, Truth table and Electrical analogue of NOT gate.

Example 2.1

Create the combination of gate for the following truth table using AND and OR gate:

A	B	C	Y
1	0	1	1
0	1	0	0
1	1	1	1
1	0	0	0
0	1	1	0
0	0	1	0

(Hint- ABC = 000 and 110 combinations are missing in table, so don't cares can be used for these combinations.)

Answer & Explanation:

$$Y = AC$$

Method 1: By analysing the table you can see that whenever A is zero, the Y is zero for sure. It means that A is getting multiplied to the other inputs to get Y. Further along with A=1 C is also '1' for all Y='1'.

$$Y = A.C$$

Method 2: Alternate way using Boolean minimisation

$Y = m_5 + m_7 + d(m_0 + m_6)$ Or **Y = AB'C + ABC = AC**

Method 3- K-map method using don't cares if needed

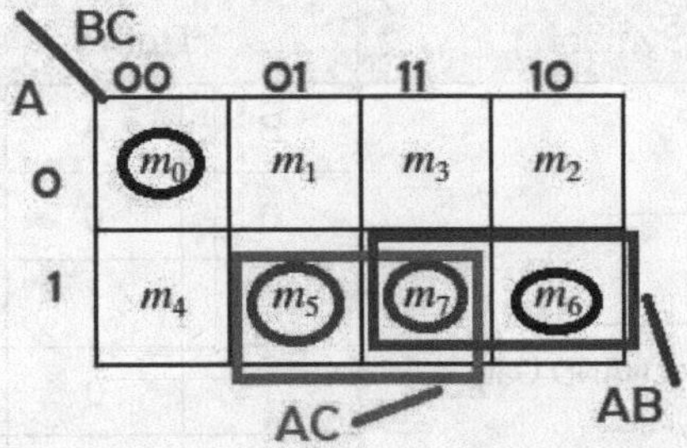

Since pair 'AC' includes both minterms, so there is no need to include don't cares. Thus minimised expression is

$$Y = AC$$

Example 2.2

Can a NAND gate be used as NOT gate?

Answer & Explanation: YES

The output of a typical 2 input NAND gate is the NOT of AND of two inputs (say A & B). So if we make both the inputs A & B same (say A). Then the output would simply be NOT of A. It can also be done by making the one input permanently to TRUE (1).

Example 2.3

Show how a NOR gate can be made into a NOT gate?

Answer & Explanation:

As explained in the above example for NAND & NOT gate, by the same logic a NOR gate can be converted into a NOT gate just by making both the inputs same or making one input permanently FALSE (0).

Electrical Analogue:

The input=0 (wire A open and A' closed) that lets current flow through bulb and it glows giving output='1'. Once the switch is turned ON (A=1), the entire current flows via the shorted circuit, while a little amount of current flows through the bulb. Output is therefore equal to '0'.

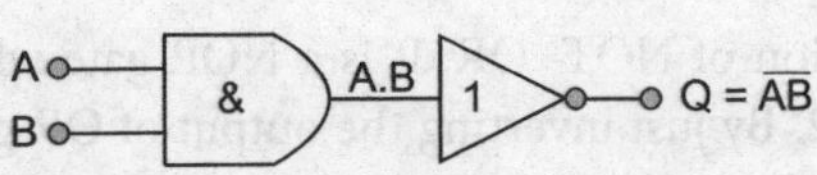

2-input "AND" gate plus a "NOT" gate

Symbol	Truth Table		
	B	A	Q
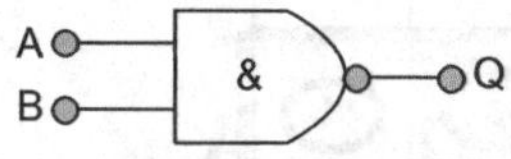 2-input NAND Gate	0	0	1
	0	1	1
	1	0	1
	1	1	0
Boolean Expression Q = $\overline{A.B}$	Read as A AND B gives NOT Q		

Fig. 2.10: NAND gate symbol and truth table

2.3.5 NAND Gate

NAND is an acronym of NOT-AND, truth table and symbolic diagrams of NAND gate are represented in Fig. 2.10.If the output entries of an AND gate's truth table are complemented, a NAND gate's truth table can be created Fig. 2.10. Here, output is a logic "0" only if every NAND gate's input is a logic "1," the else output is logic '1' in all other cases. Y = (A.B)' or Y= $\overline{A \cdot B}$ is the a NAND gate's logic expression.

An n-input NAND is represented by the Boolean equation -

$$Y = \overline{A_1 \cdot A_2 \ldots\ldots\ldots A_n}$$

Electrical Analogue:

The electrical analogue is presented in Fig. 2.11. When both switches are closed i.e. AB= 11, current flows via the short circuited channel but not through the bulb, resulting in output of '0' means bulb not shining. The bulb will glow (=> output = '1') if any switch is left open, allowing current to flow through the light bulb

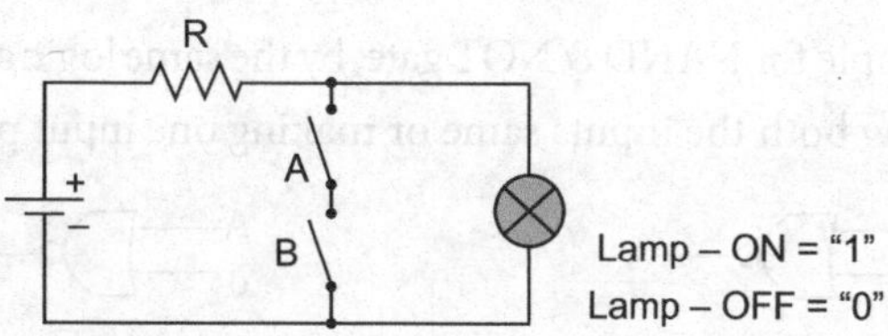

Switch A – Open = "0". Closed = "1"

Switch B – Open = "0". Closed = "1"

Fig. 2.11: Electrical analogue of NAND gate.

2.3.6 NOR Gate

NOR is a short hand representation of NOT- OR. It is a NOR gate when a NOT gate is appended to an OR gate. as shown in Fig. 2.12. By just inverting the output of OR gatetruth table of a NOR gate be created. Truth table, logic symbol for NOR gate are shown in Fig. 2.12. NOR gate gives output '1' when all of the inputs of a NOR gate are kept'0' else all other input combinations result the output is logic

'0'. Logical expression for NOR gate is represented as $Y = (A + B)'$. An n-input NOR gate can be described as -$Y = \overline{A_1 \cdot A_2 + \ldots\ldots\ldots A_n}$

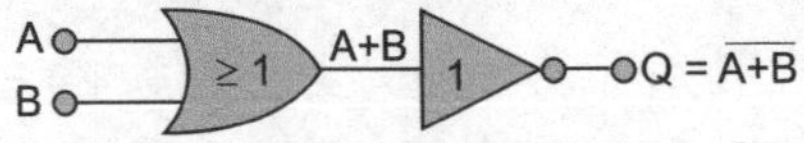

2-input "OR" gate plus a "NOT" gate

Symbol	Truth Table		
	B	A	Q
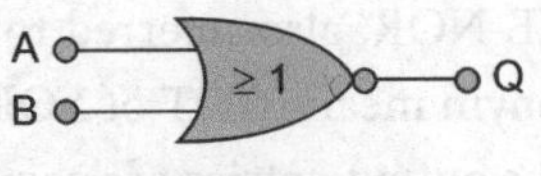 2-input NOR Gate	0	0	1
	0	1	0
	1	0	0
	1	1	0
Boolean Expression Q = $\overline{A.B}$	Read as A OR B gives NOT Q		

Fig. 2.12: NOR logic symbol and truth table of a 2 input NOR gate.

Electrical Analogue:

When both switches are open (i.e. AB=00), electricity flows through the bulb and makes output HIGH or '1' as shown in Fig. 2.13. Iff either one of both inputs are HIGH, the current will not flow through the light bulb, producing output '0'. Therefore NOR operation can be verified.

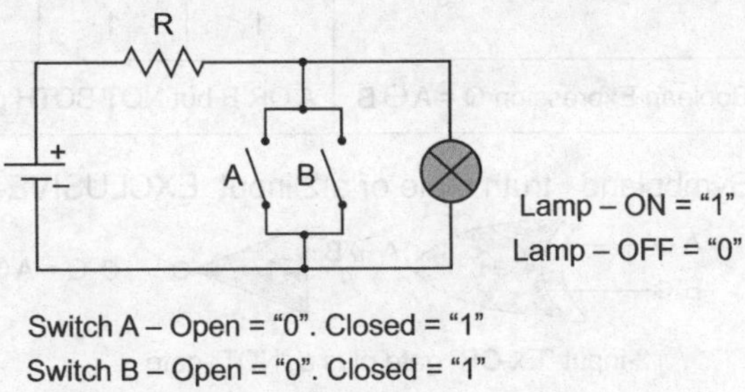

Fig. 2.13: Electrical analogue of NOR gate.

2.3.7 Exclusive-or Gate

Two inputs and one output make up the EXCLUSIVE-OR gate, sometimes called the XOR gate. The truth table and logic symbol for a two-input XOR gate are shown in Fig. 2.14. According to the truth table of two input XOR gate, an XOR gate outputs a logic '1' whenever the inputs are different from each other and a logic '0' if they are similar. XOR gates can have more than two inputs. But it is possible to construct multiple-input XOR logic functions employing more than one two-input gate. Two input XOR gate can be represented by below logic equation-

$$Y = A \oplus B$$

An n-input XOR logic gives output '1' if the number of '1' in input vector counted is odd and a logic '0' when the number of '1' in input vector counted is even. N-input XOR gate can be represented like this-

$$Y = A_1 \oplus A_2 \oplus \ldots\ldots\ldots \oplus A_n$$

Here the number of one's will be counted in the complete input sequence. For example for a 3-input XOR gate input sequence 001, 010, 100, 111 will give output HIGH since the number of 1's is either one or three in number (odd number of ones in input sequence).

2.3.8 Exclusive-nor Gate

A logic gate called EXCLUSIVE-NOR, also referred to as XNOR, is produced by complementing the result of an XOR gate. The acronym means NOT of XOR. Figure 2.15 shows its truth table as well as the logic gate symbol. An XOR gate's output entries are complemented to create truth table for XNOR gate. Logically it is represented as

$$Y = \mathbf{A} \odot \mathbf{B} \text{ or } Y = (A.B + \overline{A \cdot B}).$$

Symbol	Truth Table		
	B	A	Q
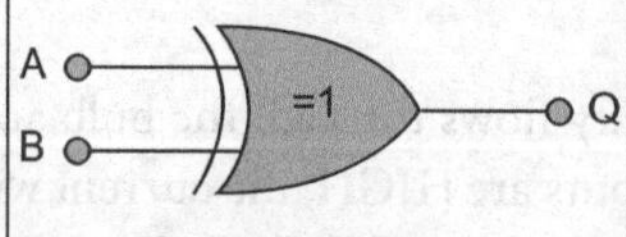 2-input Ex-OR Gate	0	0	0
	0	1	1
	0	1	1
	1	1	0
Boolean Expression $Q = A \oplus B$	A OR B but NOT BOTH gives Q		

Fig. 2.14: Symboland truth table of a 2 input EXCLUSIVE-OR (XOR) gate

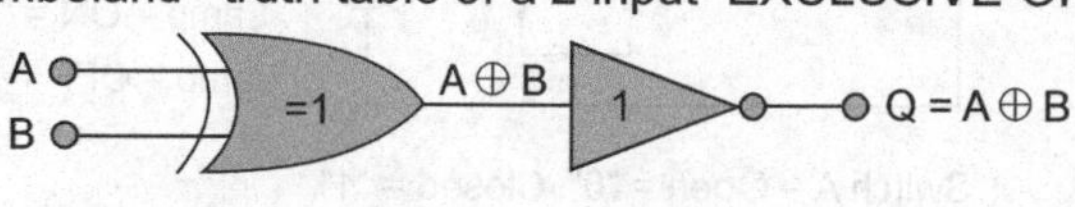

2-input "Ex-OR" gate plus a "NOT" gate

Symbol	Truth Table		
	B	A	Q
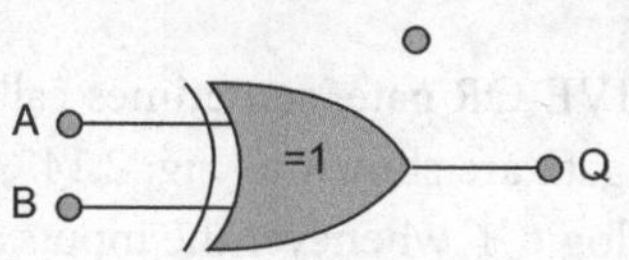 2-input Ex-OR Gate	0	0	1
	0	1	0
	1	0	0
	1	1	1
Bolean Expression $Q = \overline{A \oplus B}$	Read if A AND B the SAME give Q		

Fig. 2.15: Circuit symbol of a two-input EXCLUSIVE-NOR gate the truth table of a two-input EXCLUSIVE-NOR gate.

Example 2.4

Find one logic gate that can substitute the complete circuit after creating a Truth Table for the logical operations at the circuit's points C, D, and Q.

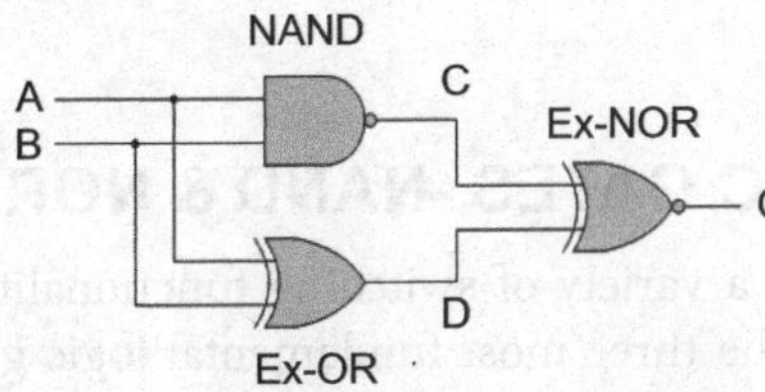

Answer & Explanation:

Since the circuit has only 2 inputs, A and B, there are only 4 possible input combinations (a total of 2^2), and these are 0-0, 0-1, 1-0, and 1-1. The truth table here, which was created by tabulating the logical functions from each gate, represents the full logic circuit.

Inputs		Output at		
A	B	C	D	Q
0	0	1	0	0
0	1	1	1	1
1	0	1	1	1
1	1	0	0	1

Example 2.5

For the given system, determine the Boolean algebraic expression.

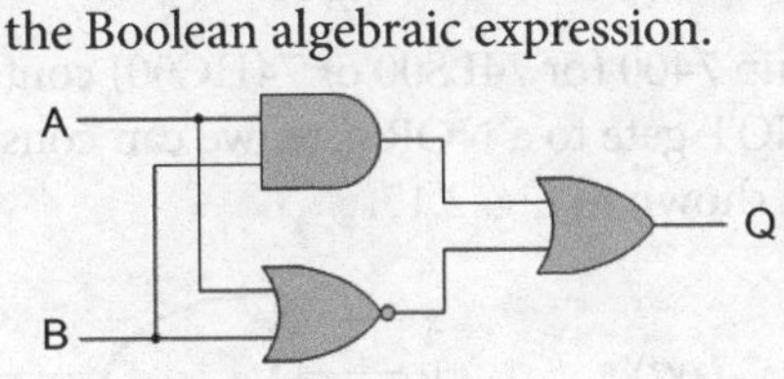

Answer & Explanation

An AND Gate, a NOR Gate, and an OR Gate make up the system. The AND gate's expression is A.B, while the NOR gate's expression is . The OR gate, which is defined as A+B, also accepts both of these expressions as distinct inputs. The following is the final output expression:

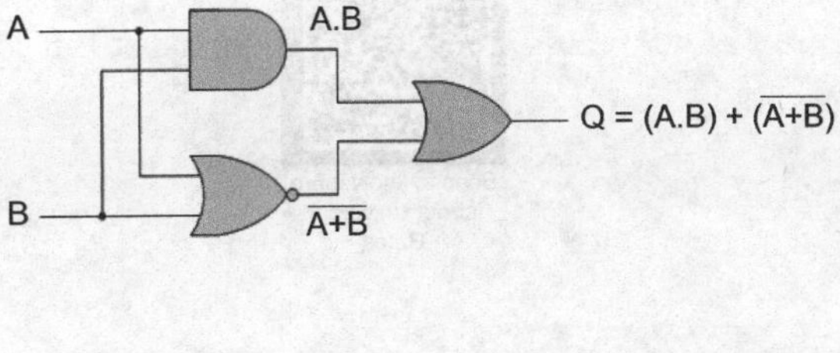

The output of a 2 inputs XNOR gate is either a logic "1" or a logic "0" depending on how similar or dissimilar the inputs are. In general, a several input XNOR logic function outputs a logic "1" when the input sequence contains an even number of 1s, including zero, and a logic "0" when the sequence contains an odd number of 1s. In other words, logic '1' is produced at the output if the input sequence is all 0s.

2.4 UNIVERSAL LOGIC GATES -NAND & NOR

Combinational logic circuits and a variety of switching functionalities can be created by connecting individual logic gates together. The three most fundamental logic gates are the AND, OR, and NOT gates. By using logical sets in this way, it is possible to apply the various rules and theorems of Boolean algebra using a whole set of logic gates. In fact, since all other Boolean functions can be created using only the set of AND and NOT gates, it is feasible to create the OR function. In the same way that OR and NOT can be joined to create the AND function.

An entire logic set is a group of gates that can be combined to realise any other logical function, whereas a universal logic gate is any logic gate that may be combined with a set to generate any other logical function. We require two (or more) different types of logic gates, NAND (AND and NOT), NOR (OR and NOT), or all 3 as indicated above, to construct any equivalent logic gate or operation when using the entire sets of AND, OR, and NOT gates.

All additional Boolean functions and gates can be implemented using the NAND (NOT AND) or NOR (NOT OR) gate, a single type of universal logic gate, thereby reducing the number of different types of logic gates required and their cost. The NAND and NOR gates are the opposites of the earlier AND and OR functions, respectively, and each one stands alone as a full set of logic because they can be combined to create some other Boolean function or gate. Because we may create further logic switching functions using only these two gates, they are known as a minimal set of gates. The NAND and NOR gates are referred to as "Universal Logic Gates."

2.4.1 Implementation using NAND Gate

The quad 2-input NAND TTL chip 7400 (or 74LS00 or 74HC00) contains four separate NAND gates in a single IC package. So, from a NOT gate to a NOR gate, we can construct all of the Boolean functions with a single 7400 TTL device, as shown in Fig. 2.17.

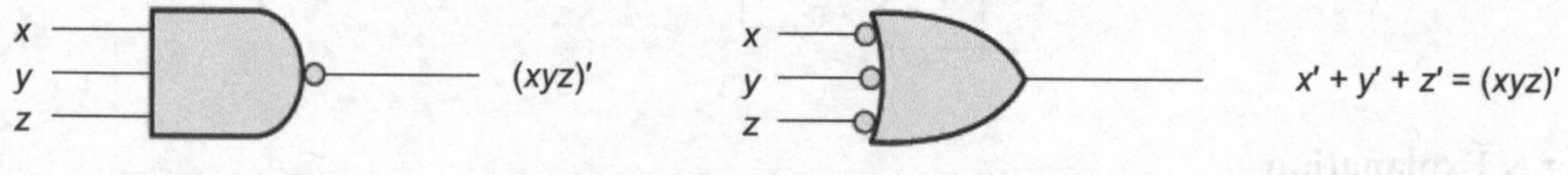

Fig. 2.16: NAND gate equivalent symbol for implementing using NAND gates

Scan to know more about Universal Gates

2.16 NAND GATE EQUIVALENT SYMBOL FOR IMPLEMENTING USING NAND GATES

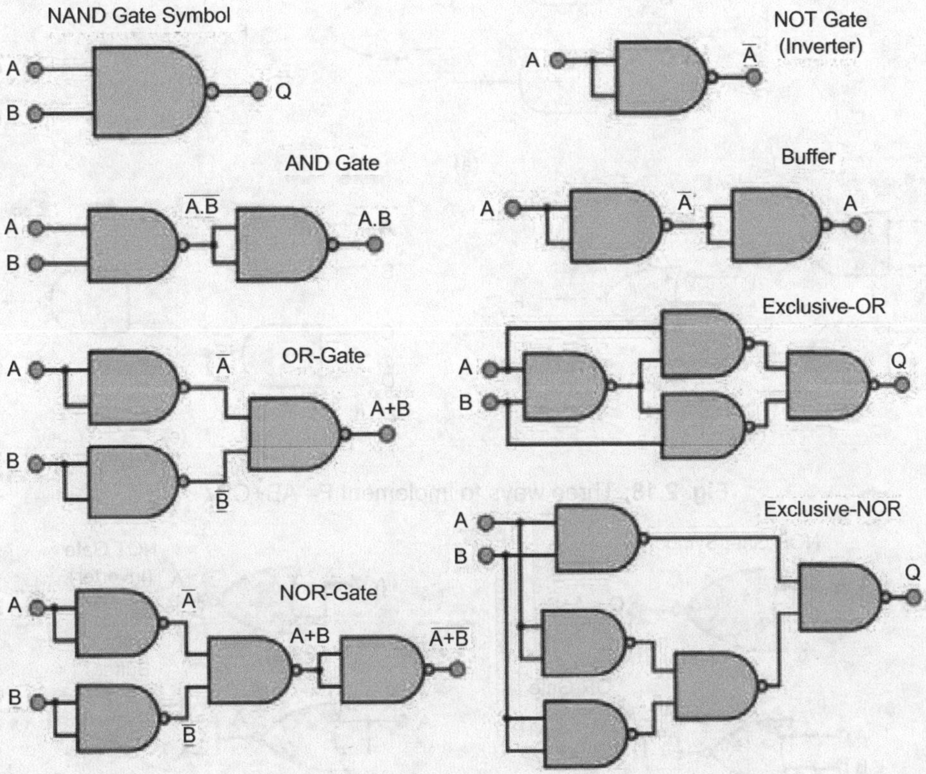

Fig. 2.17: Implementation by NAND gate of NOT gate, AND gate, Buffer, OR gate ,EX- OR gate, NOR gate, and XNOR

Another easier way to implement using NAND gates is using Demorgan's law to get another expression for NAND gate, Fig. 2.16. This will be very easy to get NAND implementation for SOP (sum -of-products) expression.

Fig. 2.17 shows that every gate can be implemented using NAND gates only. In Fig. 2.18, F=AB+CD is implemented using basic gates, NAND gate (derived symbol) and NAND gate all same symbols. Here F is given in SOP form, so, Fig. 2.18(a) can be obtained using basic gates. To implement using NAND gates, insert two NOT gates (because one circle represents one NOT gate, two circles represents a buffer) as shown in Fig. 2.18(b). Now Level -1 (gates nearest to inputs) is AND + NOT = NAND and level-2 (gates after level-1) is inverted inputs of OR gate = NAND according to Fig. 2.16. Thus Fig. 2.18(c) is derived.

2.4.2 Implementation using NOR Gate

The quad 2-input NOR TTL chip 7402 (or 74LS02 or 74HC02) contains four separate NOR gates in a single IC package. In order to produce all the Boolean functions from a single NOT gate to a NAND gate, as indicated, we can use a single 7402 TTL device,

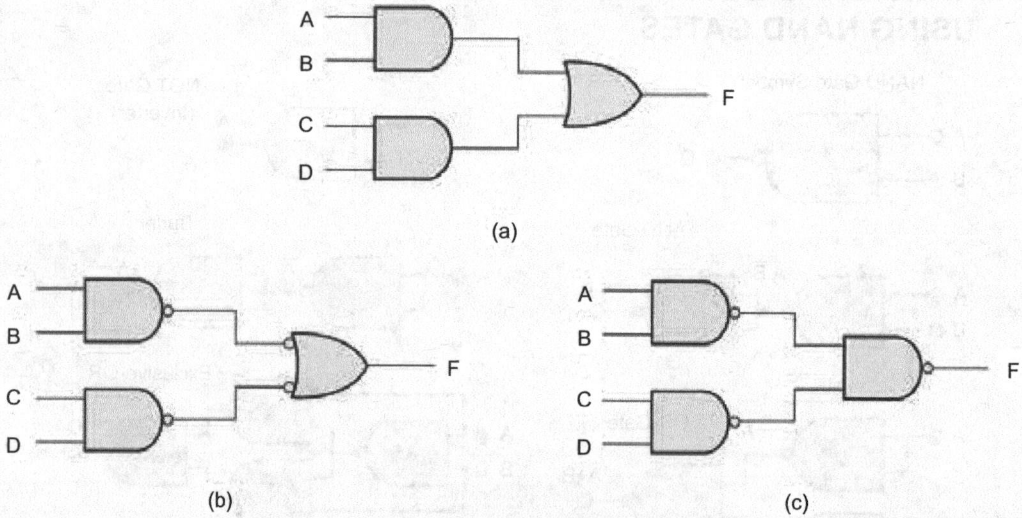

Fig. 2.18: Three ways to implement F= AB+CD

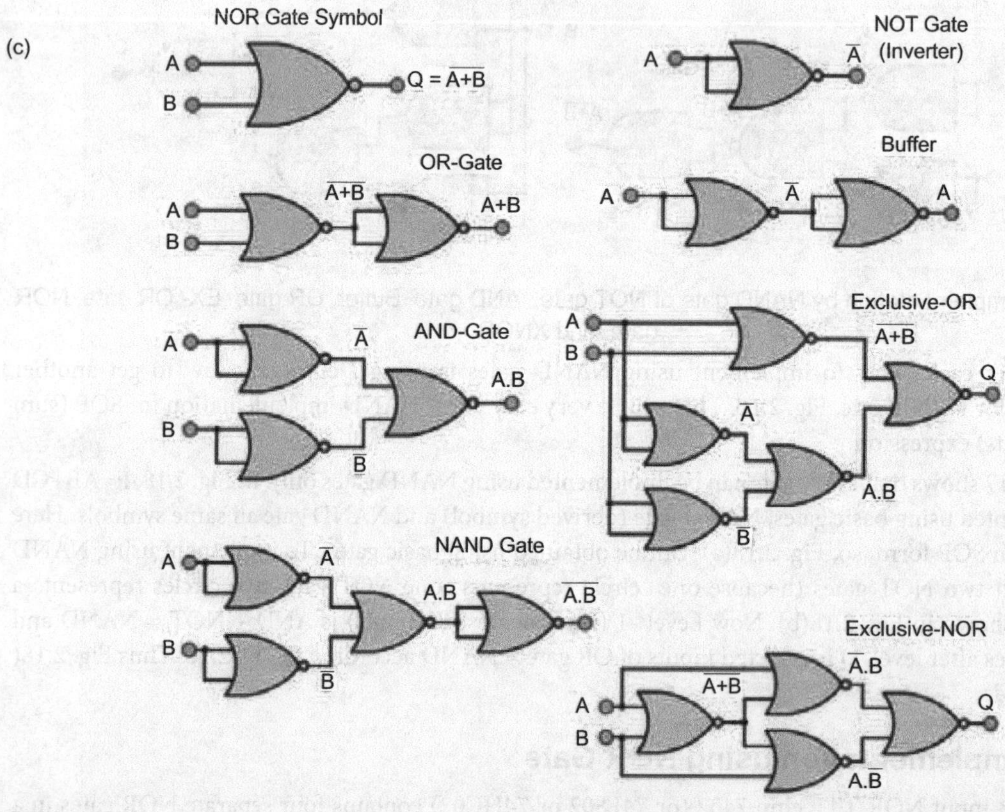

Fig. 2.19: Implementation by NOR gate of NOT gate, AND gate, Buffer, OR gate ,EX- OR gate, NOR gate, and XNOR

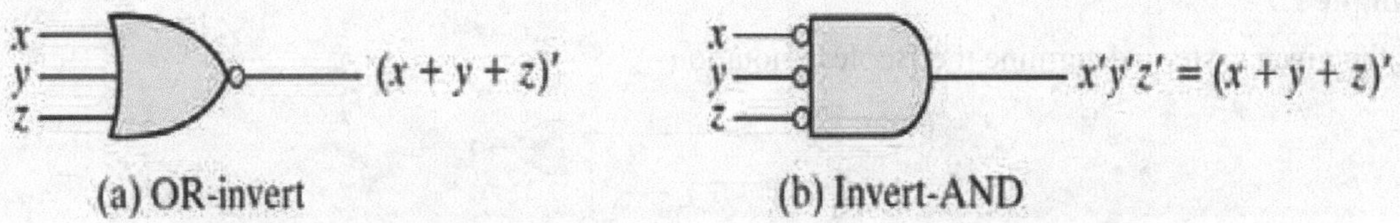

Fig. 2.20: NOR gate equivalent symbol for implementing using NOR gates

similar to the preceding 7400 NAND IC. Fig. 2.19 shows that every gate can be implemented using NOR gates only.

Another easier way to implement using NAND gates is using Demorgan's law to get another expression for NAND gate, Fig. 2.16. This will be very easy to get NAND implementation for POS (product-of-sums) expression.

Let us understand NOR implementation using F=(A+B)(C+D)E. In the below logic diagram using basic gates for F, two NOT gates (two circles= buffer gate) are inserted between level-1 and level-2. Now level-1 is NOR gate and level -2 is also NOR gate using Fig. 2.21.

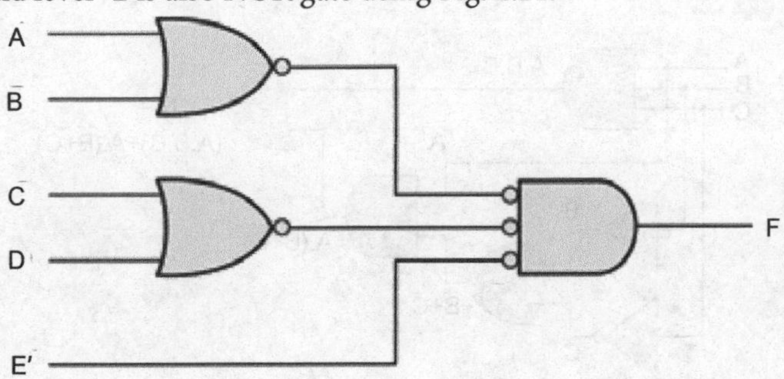

Fig. 2.21: Easier way to implement POS form using NOR gates

Example 2.6

For F= 1, except AB=11; what should be the input combination.

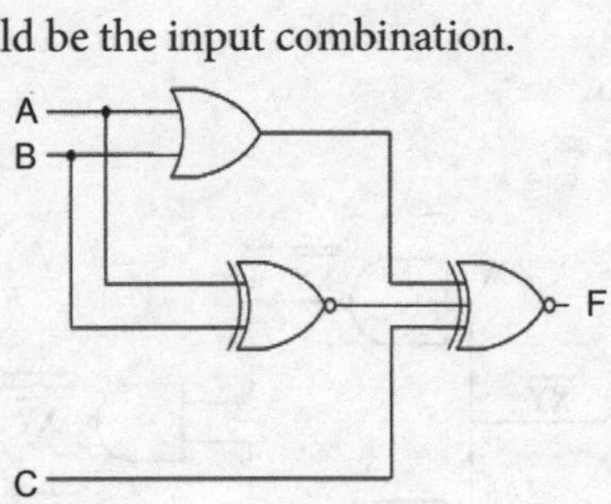

Answer & Explanation:

OR and XNOR gates are both connected to the identical inputs A and B. As a result, their output will be their complement, either 0,1 or 1,0 (wexcept AB=11). The inputs to XNOR must have an even number of 1's in order to make F = 1. Therefore, C must be '1' in order to have input 1's to be even numbers for 2nd XNOR gate.

Thus input combinations are {0X1,X01}

Example 2.7

For the given system, determine the Boolean notation.

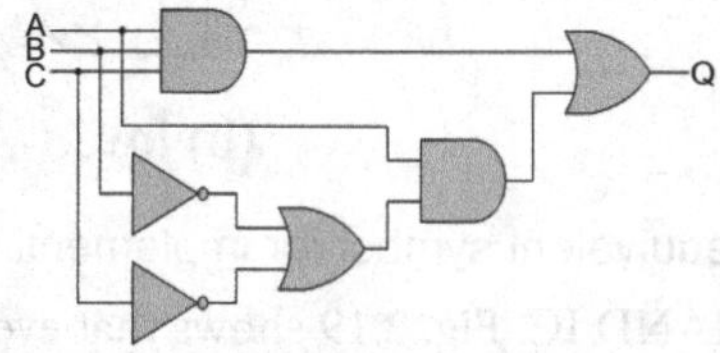

Answer & Explanation:

Only when ALL of the gate's inputs are HIGH and at logic level "1," does the three-input AND gate's output have logic "1" as its value (A.B.C). When one or both of the inputs B or C are at logic level "0," only a "1" is output from the lower OR gate. When input A is a "1" and inputs B or C are at "0," the 2-input AND gate produces a "1" as its output. If all three inputs A, B, and C are equal to "1" or if A is equal to "1" and both B and C are equal to "0," the output at Q will only be a "1" in such circumstances.

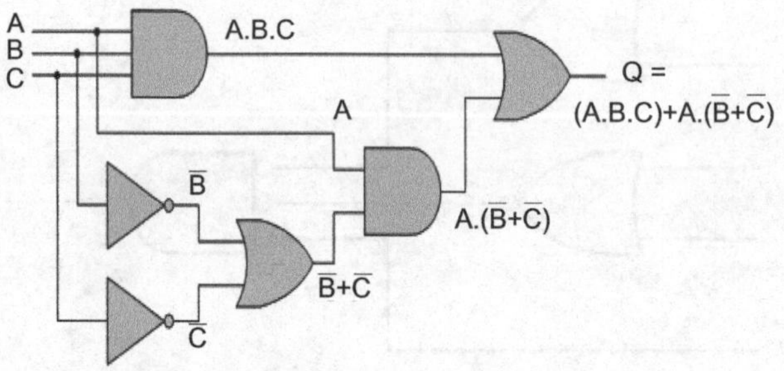

Example 2.8

Find number of gates to implement the expression:

. F = W.(X.Y)'+Z.(X.Y)'

Answer & Explanation:

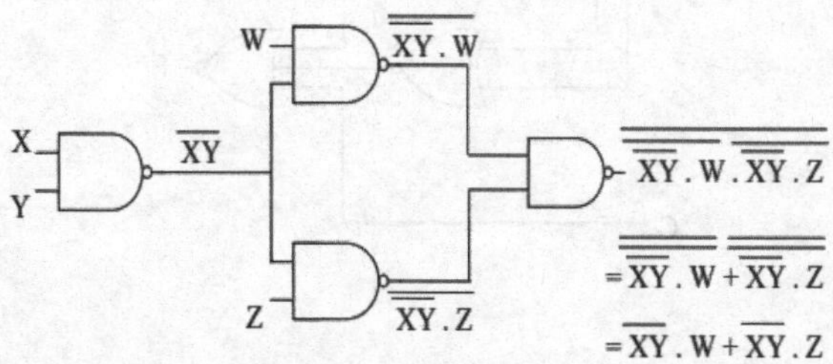

So total no of gates required is 4.

Example 2.9

The bank has rented out a locker. Explain the digital operation entailed in opening the locker.

Answer & Explanation:

One key (A), which is with the client, and the other key (B), which is with the bank, can be used to access the locker gate (F). The locker door opens when both keys are used. When both keys are used together, A = B = 1, the locker door opens, or F = 1. The procedure can therefore be described as an AND operation.

i.e. $F = A.B$

UNIT SUMMARY

In the unit, we went through the various important logic gates which are fundamental to solve any Boolean expression.

- AND, OR, NOT are basic logic gates because any circuit can be implemented using these gates.
- NAND, NOR are universal logic gates because any circuit can be implemented using these gates.
- Various important logic gates are discussed which are fundamental building blocks for any Digital Circuit.
- Implementation using NAND gates and NOR gates is explained with easier approach.

EXERCISES

Multiple Choice Questions

1. A light fixture on a staircase has two switches, one on the first floor and the other on the ground floor. Regardless of the position of the other switch, one of the switches has the ability to turn on and off the bulb. The reasoning behind altering the bulb is similar.

(a) AND gate (b) OR gate
(c) XOR gate (d) NAND gate

2. When all of a logic gate's inputs are at logic 0, the output will be "1." The gate is:

(a) NAND, EX OR (b) NOR,XNOR
(c) OR, EX NOR (d) AND, XOR

3. Boolean function f of the variables X and Y: f(0, 0) = f(0, 1) = f(1, 1) = 1; f(1, 0) = 0 Using only 2-input NOR gates and 2-input OR gates and complements of X and Y are not accessible, a least cost method for achieving would cost a total of

(a) 1 unit (b) 4 unit
(c) 3 unit (d) 2 unit

4. Assuming that A, B, and C are accessible, the bare minimum number of 2-input NAND gates needed to implement ☒ = ☒☒☒☒

(a) 2 (b) 3
(c) 5 (d) 6

5. Which of the subsequent logic gates can be utilised to realise all conceivable combinational Logic functions?

(a) OR (b) NAND-NOR
(c) EX OR (d) AND

6. Only 2 input NAND gates are needed to implement the function Y= AB + CD. The minimum number of gates is
   (a) 2 (b) 4
   (c) 3 (d) 5
7. When a NAND gate's input variables are short circuited, the result is identical to
   (a) OR (b) NAND-NOR
   (c) EX OR (d) AND
8. Which gate is not a basic gate
   (a) AND (b) NAND
   (c) NOT (d) OR
9. The number of permutations in the input side of the truth table for a 3-input logic gate is .
   (a) 2 (b) 4
   (c) 12 (d) 8
10. Gate which is represented as inverter
    (a) AND (b) NAND
    (c) NOT (d) OR
11. Suitable gate for bit comparison is
    (a) AND (b) NAND
    (c) NOT (d) EX-NOR
12. Number of basic logic gates
    (a) 2 (b) 5
    (c) 3 (d) 8
13. Gate which have 1 input only
    (a) AND (b) NAND
    (c) NOT (d) OR
14. True state shown in positive logic by
    (a) −1 (b) 1
    (c) 0 (d) −0
15. True state in negative logic shows
    (a) 1 (b) −1
    (c) Vdd (d) 0

ANSWERS															
1.	(b)	2.	(c)	3.	(c)	4.	(a)	5.	(c)	6.	(a)	7.	(b)	8.	(b)
9.	(d)	10.	(c)	11.	(d)	12.	(c)	13.	(c)	14.	(b)	15.	(d)		

Short and Long Answer Type Questions

1. Write and learn truth table for all gates with 4-inputs (except NOT gate and buffer).
2. Which are universal gates? Give instances to support your logic.
3. Create an inverter implementation utilising 2-input NAND, 2-input NOR, 2-input XOR
4. How can you differentiate a positive logic system out of a negative logic system? 4.
5. Show that a positive logic system's AND gate is equivalent to a negative logic system's OR gate.

6. What are open drain output logic gates? What are the key benefits and drawbacks of the these devices?
7. Establish the associative law for the XNOR and XOR operators.
8. Demonstrate that NOR operators for three do not fall under the associative law.
9. Demonstrate that NANDoperators for three do not fall under the associative law.
10. Describe how NOR gates can be used to implement AND, OR, and NOT gates.
11. What distinguishes IEEE/ANSI symbols from traditional ones and what is their primary meaning.
12. Create a logic diagram that uses the least amount of two-input logic gates possible to implement an 8 input XNOR function.

Numerical Problems

1. Create asix-input XORgate using the fewest possible-
   (i) NAND gates, (ii) NOR gates
2. Validate the following principles and create the corresponding circuits.
   (a) PR+PQR=PR (b) P+PQ=P
   (c) PQR+PQR+PQR =P(Q+R) (d) P+P=1
   (e) (P+Q)(P+R)=PR+PQ
3. Prepare8-input NAND gate using only fewest possible (i) two input NAND gates (ii) two input NOR gates
4. Create an inverter using (i) two-input XNOR, (ii) two-input XOR, (iii) two-input NOR and (iv) two-input NAND.
5. Implement the logic function SOP form- $F(w,x,y,z) = \Sigma(2,5,6,7,9,10,12,13,14) + d(0,3)$ using (i) Basic gates (ii) NAND gates only (iii) NOR gates only (Hint- use K-map SOP/POS form)
6. Implement the logic function POS form- $F(A,B,C,D) = \Pi(2,5,7,12,13,14,15) + d(0,3)$ using (i) Basic gates (ii) NAND gates only (iii) NOR gates only (Hint- use K-map POS form)
7. Draw logic diagrams to implement the given logic expressions:
   (a) $y = [(u + x')(y' + z)]$ (b) $y = (u'y)' + x$
   (c) $y = (u' + x')(y + z')$ (d) $y = u(x'z) + y'$
   (e) $y = u + yz + uxy$ (f) $y = u + x + x'(u + y')$

PRACTICAL

Experiments can be done in the lab using digital ICs.

- Experiments can be done on EDA Playground website using VERILOG language (Login is required using email ID).https://www.edaplayground.com
- Experiments can be performed in virtual labs as well. Here QR code for Virtual labs from IIT Bombay and IIT Kharagpur are provided respectively.

Virtual-Lab IIT Kharagpur

Virtual-Lab IIT Bombay

EXPERIMENT 1. To verify the truth tables for all logic fates – NOT, OR, AND, NAND, NOR, XOR.

Apparatus required:

Digital ICs- 7400, 7402, 7404, 7408, 7432, 7486, 747266. All ICs internal details are given in next diagram.

Procedure:

1. Place all ICs in the digital trainer kit, Each IC pin should go into a vertical connected line of beard board.
2. Connect Vcc and Gnd
3. Connect inputs to given input switches and outputs to output LEDs
4. Switch ON the trainer kit and Observe the outputs, apply all truth table input combinations
5. Switch OFF the trainer kit and compare all observations with truth table. Both must be same.

Conclusion:

All gates are verified using digital ICs.

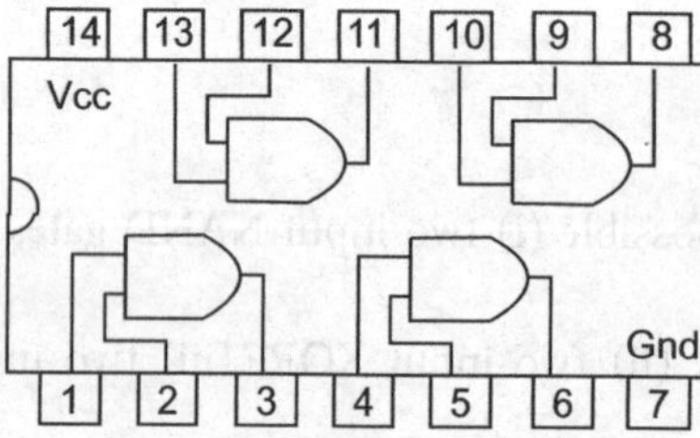

7408 Quad 2 input
AND Gates

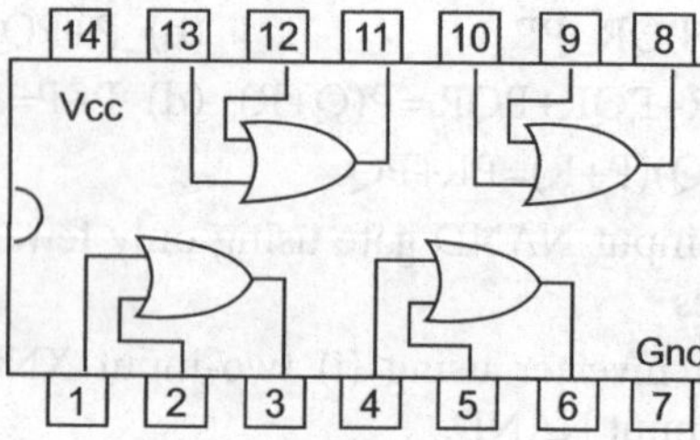

7432 Quad 2 input
OR Gates

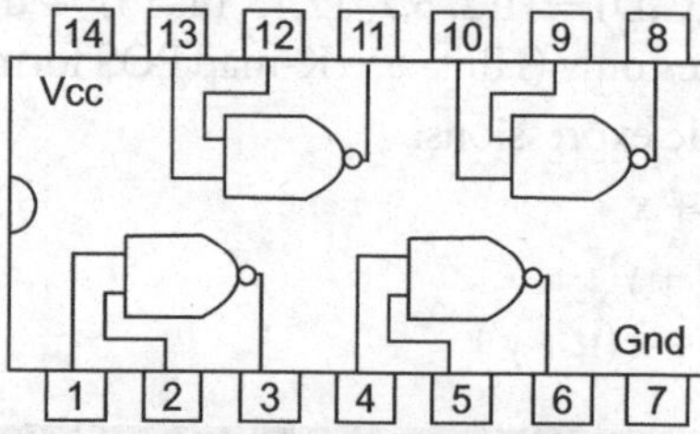

7400 Quad 2 input
NAND Gates

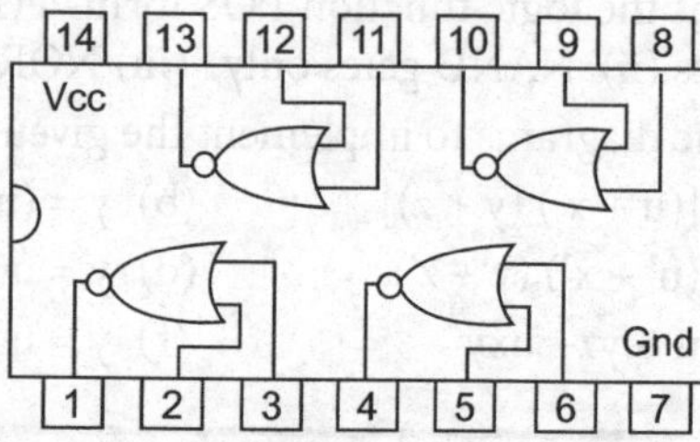

7402 Quad 2 input
NOR Gates

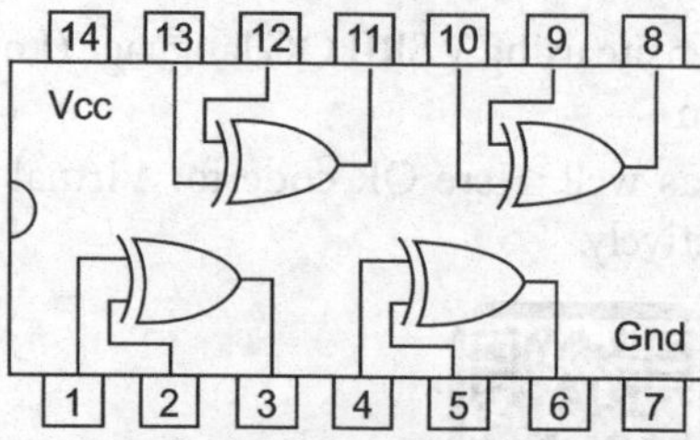

7486 Quad 2 input
XOR Gates

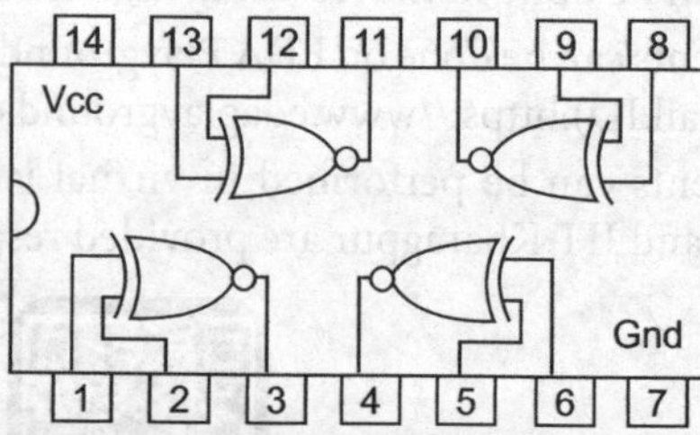

747266 Quad 2 input
XNOR Gates

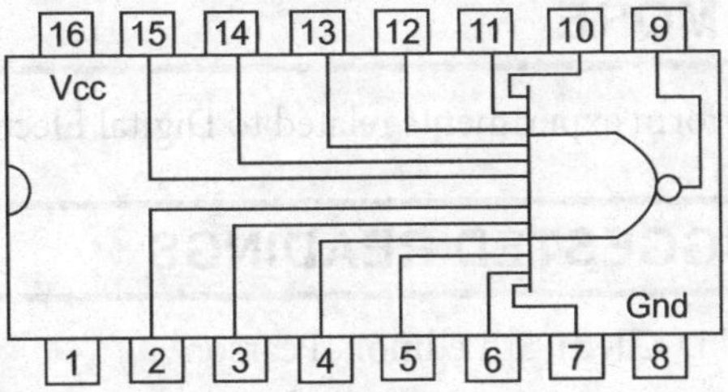

74133 Single 13 input NAND Gate

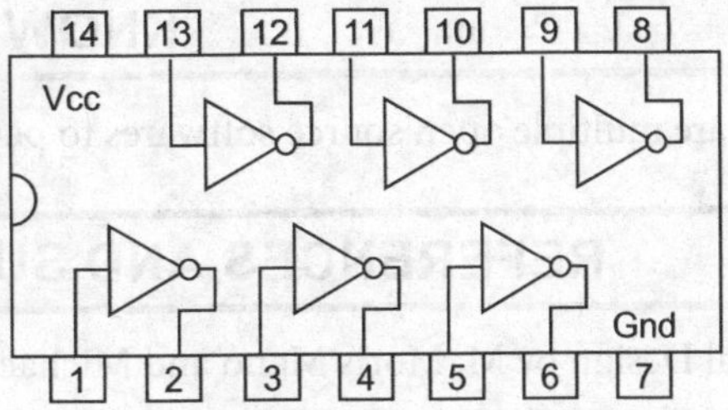

7404 Hex NOT Gates (Inverters)

** Please note that the numbering starts from bottom side of HALF CIRCLEs. Remember the ICs internal details (Input followed by output or Outputs followed by Inputs)

EXPERIMENT 2. Implement and realize Boolean Expressions with Logic Gates

F (X,Y,Z) = X+Y.Z

Apparatus: 7408, 7432 IC, jumping wires and digital trainer kit

Logic Diagram-

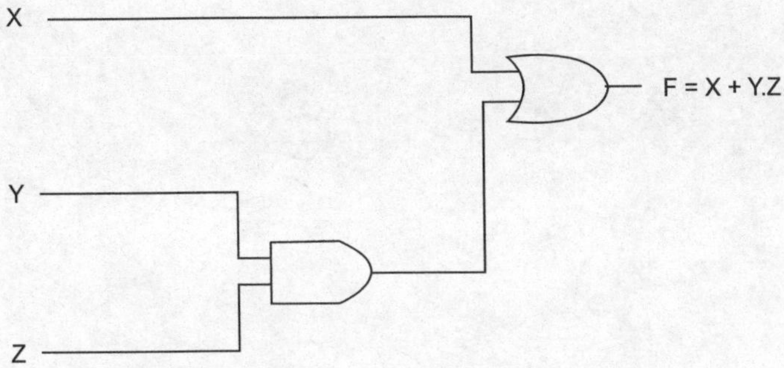

Procedure:

1. Place both ICs in the digital trainer kit,
2. Connect Vcc and Gnd
3. Connect inputs Y and Z to AND gate IC (7408). Connect Output of AND gate to Input of OR gate IC 7432 and another input to X input. Connect output OR gate to output LED
4. Switch ON the trainer kit. Apply all truth table input combinations and observe the outputs.
5. Switch OFF the trainer kit and compare all observations with truth table. Both must be same.

Conclusion:

Given Boolean expression is verified using Digital ICs.

- The same experiments (1 & 2) can be performed using virtual labs at the below link.

  Virtual Labs (iitkgp.ac.in)

KNOW MORE

There are multiple open source softwares to perform experiments related to Digital Electronics.

REFERENCES AND SUGGESTED READINGS

- Digital Design by M. Moris Mano and Michael D Ciletti, 5th edition, Pearson
- Digital Electronic Circuits - Course (nptel.ac.in)

Dynamic QR Code for Further Reading

- QR codes embedded in the chapter

□□

3 Combinational Logic Circuits

UNIT SPECIFICS

Through this unit we have discussed the following aspects:

- Addition and Subtraction of Arithmetic Circuits;
- Addition and Subtraction using "1's and 2's complement";
- Designing of "Half Adder", "Full Adder";
- Designing of "Half Subtractor", "Full Subtractor";
- "Parallel Adder" and "Series Adder";
- "Encoder", "Decoder";
- "Multiplexer", "Demultiplexer" and their applications.

The practical applications of the topics are discussed for generating further curiosity and creativity as well as improving problem solving capacity.

Besides giving a large number of multiple choice questions as well as questions of short and long answer types marked in two categories following lower and higher order of Bloom's taxonomy, assignments through a number of numerical problems, a list of references and suggested readings are given in the unit so that one can go through them for practice. It is important to note that for getting more information on various topics of interest some QR codes have been provided in different sections which can be scanned for relevant supportive knowledge.

After the related practical, based on the content, there is a "Know More" section. This section has been carefully designed so that the supplementary information provided in this part becomes beneficial for the users of the book. This section mainly highlights the initial activity, examples of some interesting facts, analogy, history of the development of the subject focusing the salient observations and finding, timelines starting from the development of the concerned topics up to the recent time, applications of the subject matter for our day-to-day real life or/and industrial applications on variety of aspects, case study related to environmental, sustainability, social and ethical issues whichever applicable, and finally inquisitiveness and curiosity topics of the unit.

RATIONALE

This fundamental unit helps students to get a primary idea about the designing of combinational circuits using logic gates. It explains about the designing of arithmetic circuits. It also explains about the "1's and 2's complement" of "binary numbers" and to "perform the arithmetic operations using these numbers".

All these basic aspects are relevant to start the learning of combinational circuits properly. The chapter also explains about the converter circuits such as encoder, decoder and other combinational circuits like Multiplexer, Demultiplexer which are used for data transmission. Some related problems are pointed regarding the designing of combinational logic circuits which can help further for getting a clear idea of the concern topics on digital electronics.

As the scope of digital electronics extend to domains such as healthcare, software, automation, digital banking and many more, various combination of logic gates are utilized to represent logic signal, known as Combinational logic circuits. These are important part of digital electronics that are used in wide variety of applications such as computers, digital communication, calculators, smart devices, automatic control of machines.

PRE-REQUISITES

Basic concept of logic gates, K-map, Boolean logic

UNIT OUTCOMES

List of outcomes of this unit is as follows:

U3-O1: Understand arithmetic operations of digital numbers

U3-O2: Learn Series and Parallel Adders

U3-O3: Understand the operation of using Encoder and Decoder

U3-O4: Learn the operation of Multiplexer and Demultiplexer

U3-O5: Designing the Combinational circuits using logic gates

Unit-3 Outcomes	EXPECTED MAPPING WITH COURSE OUTCOMES (1 - Weak Correlation; 2 - Medium Correlation; 3 - Strong Correlation)					
	CO-1	CO-2	CO-3	CO-4	CO-5	CO-6
U3-O1	3	3	2	-	-	-
U3-O2	2	3	3	3	-	-
U3-O3	-	3	3	-	-	-
U3-O4	-	3	3	-	-	-
U3-O5	-	3	3	-	-	-

"When you talk, you are only repeating something you know. But if you listen, you may learn something new."

– Dalai Lama

3.1 INTRODUCTION

A combinational circuit is the type of digital logic in which "output depends only on the present inputs combination". The logic gates are the building blocks for designing a combinational circuit. The logic functions implemented by combinational circuits are in the form of Boolean expressions. Combinational circuits may have multiple outputs or single output. Figure 3.1 shows the general "block representation" of a "combinational circuit" with "m input variables and n output variables".

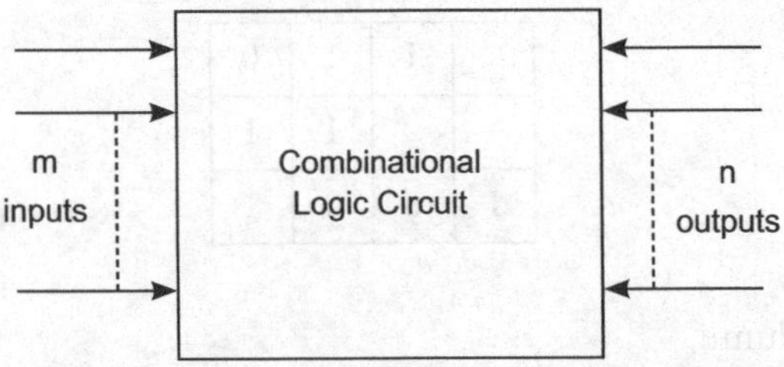

Fig. 3.1: "Block diagram" of "Combinational logic circuit" with "m inputs and n outputs"

3.2 BINARY NUMBERS: ADDITION AND SUBTRACTION

The arithmetic operations of "binary numbers" using 1's/2's complement are discussed in this chapter, and also the various types of combinational logic blocks that performs addition and subtraction operations such as "half adder, full adder, half subtractor and full subtractor" are discussed.

3.2.1 Binary Addition and Subtraction: General Rules

"Binary addition" rules-

- Adding two bits gives "0 sum" if both bits are same otherwise sum is "1" if bits are different
- Carry is "1" only when both bits are "1"

"Binary subtraction" rules-

- Subtracting two bits gives "0 difference" if both bits are same otherwise difference is "1" if bits are different

Borrow is "1" only when first bit (A) is "0" and next bit (B) is "1" from operation "A-B"

3.2.2 Addition/Subtraction using 1's Complement

There can be following three cases for Addition or subtraction using 1's complements-

Case 1: To Add positive number with a negative number (negative number has less value than positive number)

Write the "1's complement" of "negative number"

Add this number to positive number

If there is carry '1', add this to the LSB

Example 3.1

Add the numbers $(110)_2$ and $(11)_2$

Solution:

	1	1	0
+		1	1
1	0	0	1

1

	1	1	0
+		1	1
1	0	0	1

Explanation:

Start with the rightmost column,

addition 0 + 1 = 1 and rightmost digit answer is '1'

Move to next column to the left,

addition 1 + 1 = 0 and carry '1' that goes to next column

Move to next column to the left,

addition previous carry '1' + 1 = 0 and carry '1'

The final answer (from left to right) = $(1001)_2$

Example 3.2

Subtract the numbers $(101)_2$ and $(11)_2$

Solution:

	1	0	1
–		1	1
	0	1	0

	1 (borrow)		
	1	0	1
–		1	1
	0	1	0

Explanation:

Start with the rightmost column,

subtract 1-1=0 and rightmost digit answer is ***'0'***

Move to next column to the left,

*subtract 0-1=**1** and borrow '1' that goes to next column*

Move to next column to the left,

*subtract previous borrow '1'-1=**0***

The final answer (from left to right)=$(010)_2$

Case 2: To Add positive number with a negative number (negative number has more value than positive number)

- Write the "1's complement" of "negative number"
- Add this number to positive number
- The result will be "– (1's complement of the addition)"

Example 3.3

Calculate $(12)_{10}$-$(4)_{10}$ using 1's complement.

Solution:

Binary representation of $(12)_{10} = (1100)_2$
Binary representation of -$(4)_{10} = (1011)_2$

	1	1	0	0
+	1	0	1	1
1	0	1	1	1

Answer:

$$0111+1 = (1000)_2$$

Explanation:

Write the 1's complement of negative number-

4 = 0100; –4 = (1's complement of 4) = 1011

Add 1011 to 1100

If there is carry '1', add this to Least Significant Bit (LSB).

Case 3: Both numbers are negative

- Write the 1's complement of both the numbers with one extra MSB
- Add both the numbers
- Add carry to the Least Significant bit and the result will be "–(1's complement of the addition)"

3.2.3 Addition/Subtraction using 2's complement

There can be following three cases for Addition or subtraction using 1's complements-

Case 1: To Add "positive number" with a "negative number" ("negative number" has less value than "positive number")

- Write the 2's complement of negative number
- Add this number to positive number
- If there is carry '1', discard it

Case 2: To Add "positive number" with a "negative number" ("negative number" has more value than "positive number")

- Write the 2's complement of negative number
- Add this number to positive number
- The result will be "– (2's complement of the addition)"

Example 3.4

Calculate $(12)_{10} - (15)_{10}$ using 1's complement.

Solution:

Binary representation of $(12)_{10} = (1100)_2$

Binary representation of $-(15)_{10} = (0000)_2$

	1	1	0	0
+	0	0	0	0
	1	1	0	0

Answer:

("1's complement of 1100")= - $(0011)_2$

Explanation:

Write the 1's complement of negative number-

15 = 1111; –15 = (1's complement of 15) = 0000

Add 0000 to 1100

The result will be "– (1's complement of the addition)"

Example 3.5

Calculate $-(8)_{10} - (6)_{10}$ using 1's complement.

Solution:

Binary representation of $-(8)_{10} = (10111)_2$

Binary representation of $-(6)_{10} = (11001)_2$

	1	**1**	**1**	**1**	
	1	0	1	1	1
+	1	1	0	0	1
1	1	0	0	0	0

10000 + 1 = 10001

Answer: -(1's complement of 10001)= – $(01110)_2$

Explanation:

Write the 1's complement of negative number (with extra bit at MSB)-

8=01000; -8=(1's complement of 8)=10111

6=00110; -6=(1's complement of 8)=11001

Add 10111 and 11001

Add carry to LSB

The result will be "– (1's complement of the addition)"

Important Points (1's complement addition/subtraction):

–One number is positive and another number is negative:

- If Carry is there, add it to LSB
- If no carry, the result will be "- (1's complement of the result)"

-Both numbers are negative-

- Add carry to LSB and result will be "- (1's complement of the result)"

Example 3.6

Calculate $(14)_{10} - (10)_{10}$ using 2's complement.

Solution:

Binary representation of $(14)_{10} = (1110)_2$

Binary representation of $-(10)_{10} = (0110)_2$

	1	1		
	0	1	1	0
+	1	1	1	0
1	0	1	0	0

Answer: $(0100)_2$

Explanation:

Write the 2's complement of negative number-

10 = 1010; –10 = (1's complement of 10 + 1) = 0101 + 1 = 0110

Add 0110 to 1110

If there is carry '1', discard it

Case 3: Both numbers are negative

- Write the ***2's complement of both the numbers with one extra MSB***
- ***Add both the numbers***
- ***Add ca***rry to the LSB and final result will be "–(2's complement of the addition result)"

Important Points (2's complement addition/subtraction):

One number is "positive" and another number is "negative":

- If Carry is there, discard it
- If no carry, the result will be "- (2's complement of the result)"

Both numbers are negative-

- Discard carry and result will be "- (2's complement of the result)"

Example 3.7

Calculate $(9)_{10}$-$(15)_{10}$ using 2's complement.

Solution:

Binary representation of $(9)_{10} = (1001)_2$

Binary representation of $-(15)_{10} = (0001)_2$

	1			
	1	0	0	1
+	0	0	0	1
	1	0	1	0

Answer:

- (2's complement of 1010)= – $(0110)_2$

Explanation:

Write the 2's complement of negative number-

15 = 1111; –15 = (1's complement of 15 + 1) = 0000 + 1 = 0001

Add 0001 to 1100

The result will be "– (2's complement of the addition)"

Example 3.8

Calculate $-(10)_{10} - (11)_{10}$ using "2's complement".

Solution:

Binary value of

$-(10)_{10}=(10110)_2$ and

$-(11)_{10}=(10101)_2$

	1				
	1	0	1	1	0
+	1	0	1	0	1
1	0	1	0	1	1

Answer:

- (2's complement of 01011) = – $(10101)_2$

Explanation:

Write the 2's complement of negative number (with extra bit at MSB)-

10 = 01010; –10 = (2's complement of 10) = 01010 + 1 = 10110

11 = 01011; –11 = (2's complement of 11) = 10100 + 1 = 10101

Add 10110 and 10101

Discard carry

The result will be "– (2's complement of the addition)"

3.3 ARITHMETIC CIRCUITS

"Arithmetic circuits" are the "basic building blocks" of combinational circuits. These includes "half adder", "full adder", "half subtractor", "full subtractor", "parallel adder" and "series adder".

3.3.1 Half Adder

A "half adder" is a type of "arithmetic logic circuit" that "adds two binary bits". It produces the results "Sum" (S) and "Carry" (C) at the output. The "truth table" and "block diagram" are given in figure 3.2. The logic expressions of "Sum" (S), "Carry" (C) are given below-

$$S = B \oplus A = \overline{A}.B + A.\overline{B}$$

$$C = B.A$$

Inputs	Outputs	
AB	C	S
00	0	0
01	1	1
10	1	1
11	1	0

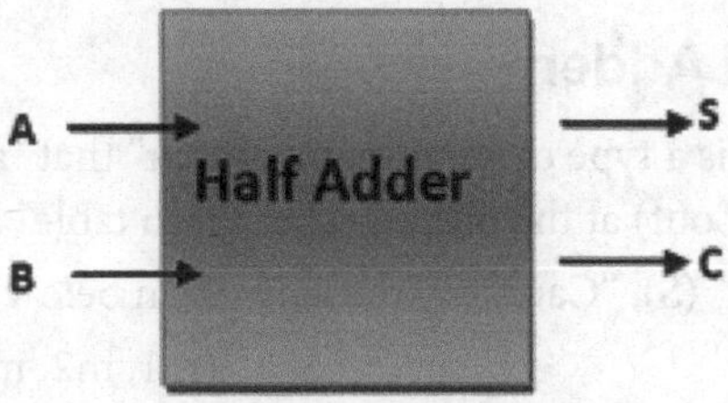

Fig. 3.2: Truth table and "block diagram" of "Half adder"

The logic gate implementation of "half adder" is given below in figure 3.3. It is the combination of a "XOR" gate for sum and an "AND" gate for carry outputs.

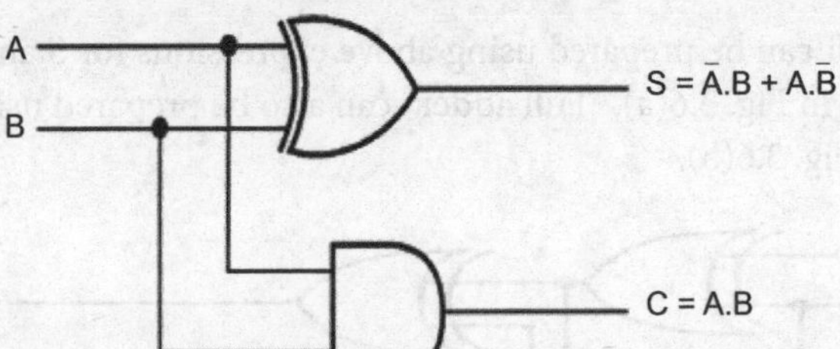

Fig. 3.3: Logic gate implementation of Half adder

Inputs	Outputs	
ABC_{in}	C_{out}	S
000	0	0
001	0	1
010	0	1
011	1	0
100	0	1
101	1	0
110	1	0
111	1	1

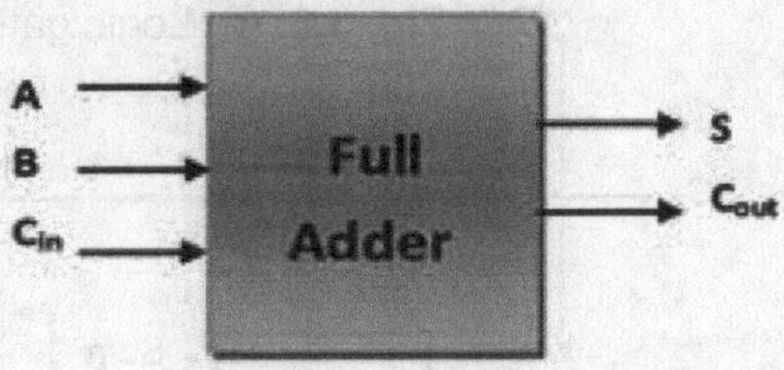

Fig. 3.4: "Truth table" and "block diagram" of "Full adder"

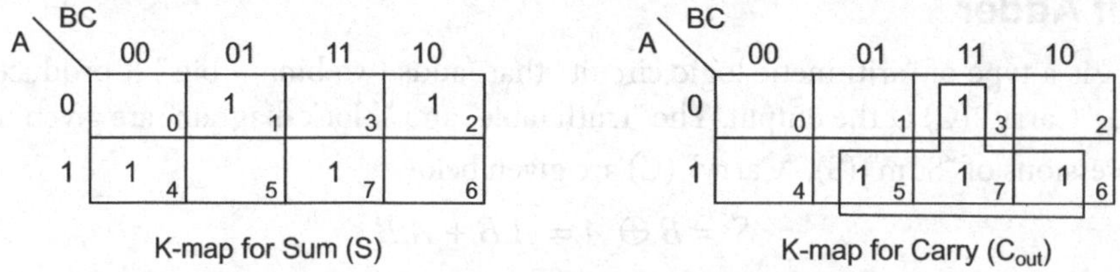

Fig. 3.5: "K-map minimisation" for "Sum" and "Carry" for "Full Adder"

3.3.2 Full Adder

A "full adder" is a type of "arithmetic circuit" that "adds three binary bits". It produces results "Sum" (S) and "Carry" (Cout) at the output. The "truth table" and "block diagram" are given in Fig. 3.4.

The "Sum" (S), "Carry" (Cout) are given below using truth table given in Fig. 3.4.-

$$S = \Sigma(m1, m2, m4, m7)$$

$$Cout = \Sigma(m3, m5, m6, m7)$$

The simplified expression of S and Cout can also be written using K-map minimisation as shown in Fig. 3.5, as-

$$S = Cin \oplus B \oplus A$$

$$C_{out} = C_{in}(B + A) + A.B$$

Full adder logic circuit can be prepared using above expressions for S and Cout. The Logic circuit for 1-bit full adder is given in Fig. 3.6(a). "Full adder" can also be prepared using "two -half adders" and an "OR gate" as shown in Fig. 3.6(b).

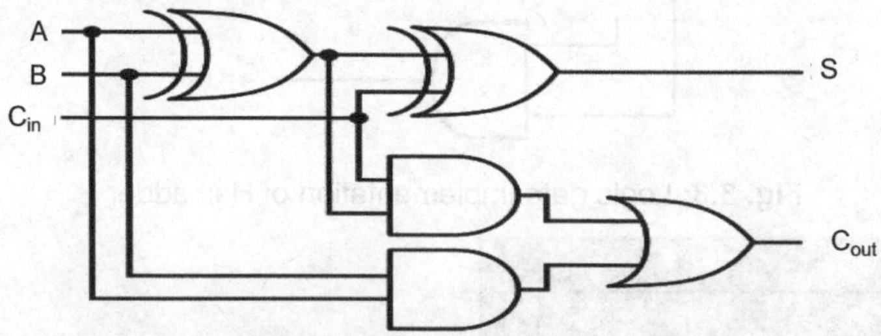

Fig. 3.6: (a) Logic gate implementation of Full adder

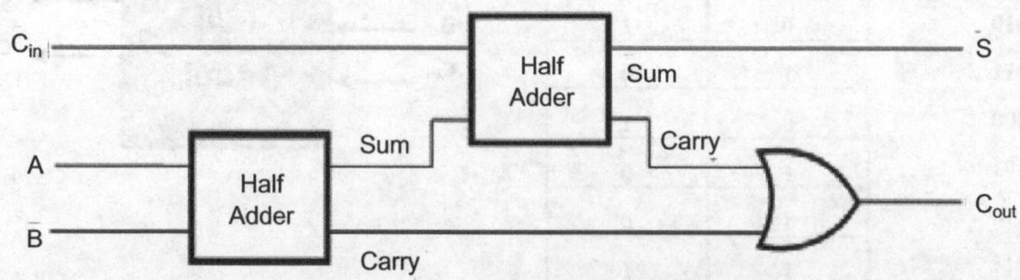

Fig. 3.6: (b) Full adder implementation using half adders

Input	Output	
AB	B_0	D
00	0	0
01	1	1
10	0	1
11	0	0

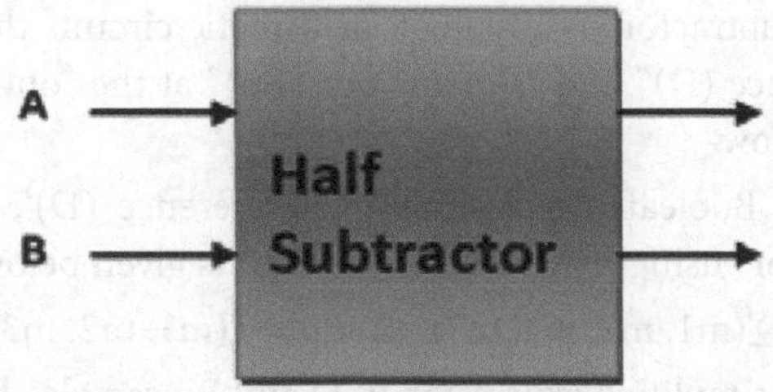

Fig. 3.7: "Truth table" and "bock diagram" of "half subtractor"

3.3.3 Half Subtractor

A "half subtractor" is a type of "arithmetic circuit" that "subtracts" "two-binary bits". It produces results "Difference"(D) and "Borrow"out (Bo) at the output. Here 'Borrow' means that '1' is borrowed to perform subtraction operation. If the difference is '-1', then 'Borrow (Bo)' will be '1'. The truth table and block diagram are given in fig. 3.8.

The logic expressions for 'D' and 'Bo' are given below-

$$D = B \oplus A$$

$$B_o = B.\bar{A}$$

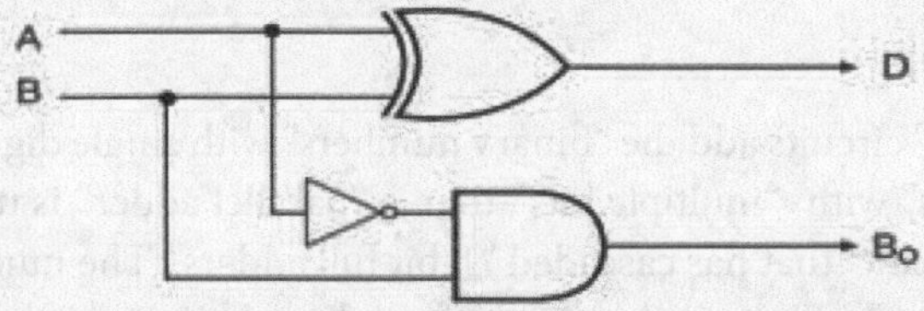

Fig. 3.8: "Logic gate implementation" of "half subtractor"

Table 3.1: "Truth Table" of "Full Subtractor"

Inputs	Outputs	
ABB_{in}	B_o	D
000	0	0
001	1	1
010	1	1
011	1	0
100	0	1
101	0	0
110	0	0
111	1	

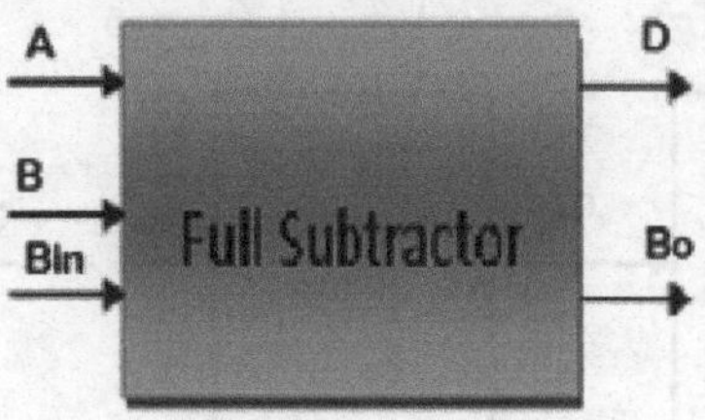

Fig. 3.9: "Block Diagram" of "Full Subtractor"

Encoder: Scan for explanation

3.3.4 Full Subtractor

A "full subtractor" is a type of "arithmetic circuit" that subtracts "three binary bits". It produces results "Difference (D)" and "Borrow out (Bo)" at the "output". The block representation and truth table are given below-

The Boolean expressions of "Difference (D)", "Borrow out (Bo)" and implementation of "full subtractor" using "two half subtractors" is given below -

D = Σ(m1, m2, m4, m7) & B0 = Σ(m1, m2, m3, m7)

The simplified expression of D and Bo can also be represented as shown in Fig. 3.10.

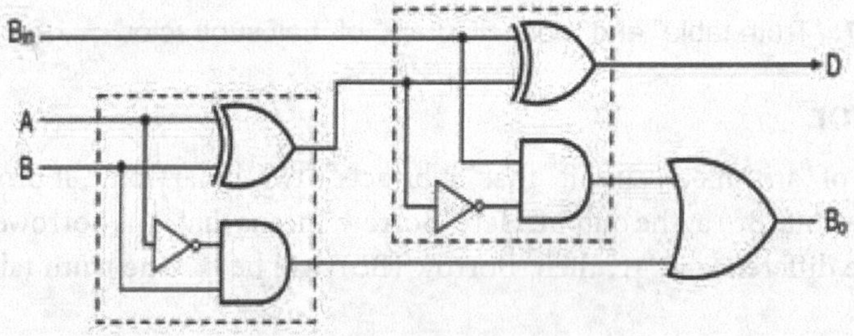

Parallel Adder: Scan for explanation

Fig. 3.10: Implementation of "full subtractor" using "two half subtractors"

"Full subtractor" can be designed using "half subtractors" as shown in figure 3.10. It is the "combination" of "two half subtractors" and one "OR gate."

3.3.5 Parallel Adder

"Half adder" and "full adder" circuits add the "binary numbers" with single digits in them, but if we want to add "two binary numbers" with "multiple bits", then a "parallel adder" is used. A "Parallel adder" is a type of "combinational circuit" that has cascaded "1-bit full adders". The number of "full adders" used to implement a "parallel adder" will depend upon the "number of binary bits" required to be added. Fig. 3.11 shows an "n-bit parallel adder" formed by cascading "n-full adders".

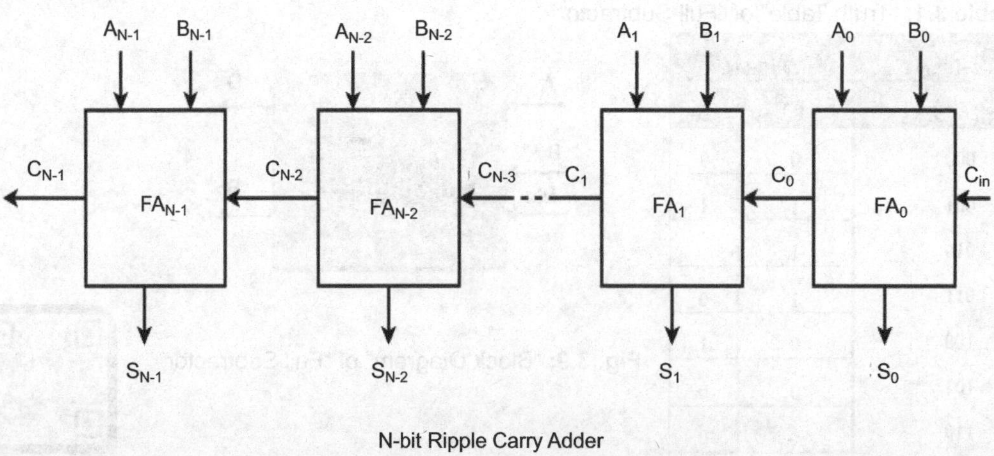

Fig. 3.11: n-bit parallel adder circuit

In the diagram shown above, first, FA0 adds A0, B0 and Cin to produce Sum (S0) and Carry (C0). Next, FA1 uses this C0 as carry-in bit and adds it with the inputs A1, B1 to produce Sum (S1) and Carry (C1). This process continues till the (N-1)th full adder and the final result is Sum (SN-1) and Carry (CN-1).

3.3.6 Series Adder

"Serial adder" or "series adder" is a type of "combinational circuit" that adds "two binary numbers" in "serial form" means one after other. It uses "two shift registers" to store the "binary numbers" and performs bit by bit addition. It consists of "two shift registers" that stores the "binary numbers", a "full adder" that takes "three inputs" and provides "two outputs", a "D-flip-flop" that is used as a "carry input" for the "full adder".

The full adder adds one pair of bits along with carry. The "D-flip-flop" gets the "carry output" from the "full adder" and this "carry output" is used as an "input carry" for "the next significant bit". Fig. 3.12 shows the "block diagram" of "serial adder".

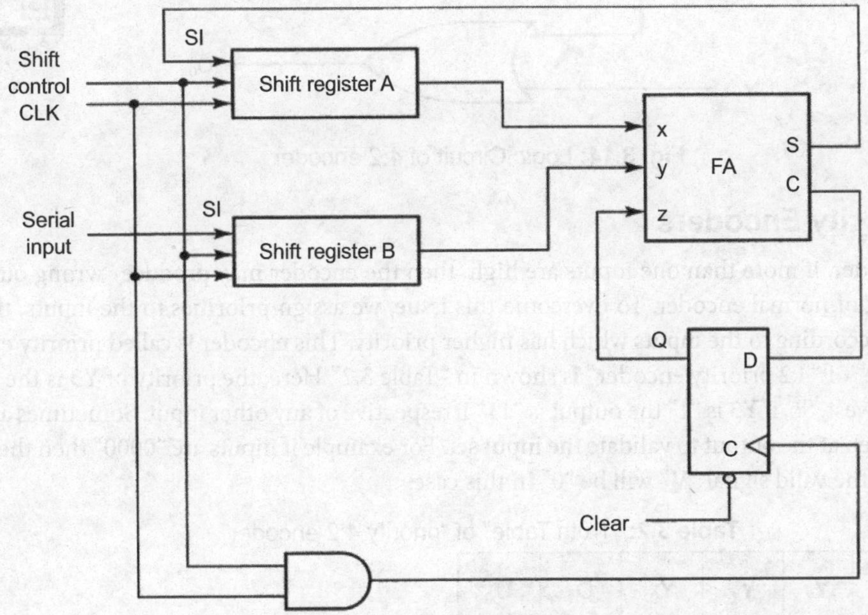

Priority Encoder: Scan for explanation

Fig. 3.12: Serial adder circuit

3.4 ENCODERS

An "encoder" is a type of "combinational circuit" that produces the binary equivalent to the input. It has "2n input lines" and "n output lines". It takes the binary information in the form of "2n input lines" and defines "n-bit code" for the binary information at the output. It can be represented as 2n : n encoder. For simplicity, only one input is activated at a time. The "block diagram" and "truth table" of "4:2 encoder" is shown in fig. 3.13.

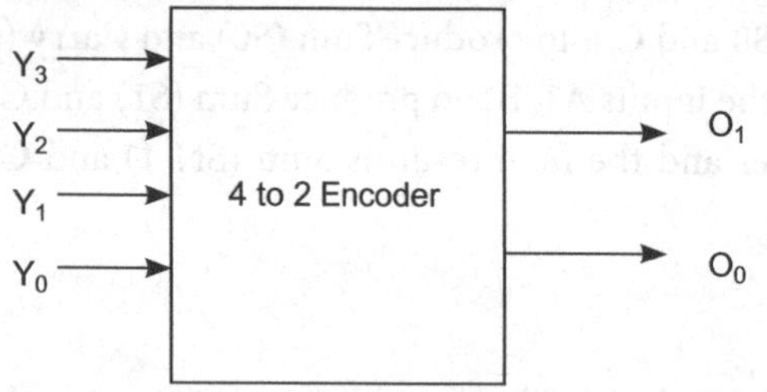

Y_3	Y_2	Y_1	Y_0	O_1	O_0
0	0	0	1	0	0
0	0	1	0	0	1
0	1	0	0	1	0
1	0	0	0	1	1
1	1	1	0	0	0

Fig. 3.13: "Block diagram" and "truth table" of 4:2 encoder

From the "truth table", the "Boolean expression" for the encoder are given as-

O1 = Y2 + Y3, O0= Y 1+Y3

The "logic circuit" of the "4:2 encoder" is given below in fig. 3.14. It can be implemented using two "OR" gates.

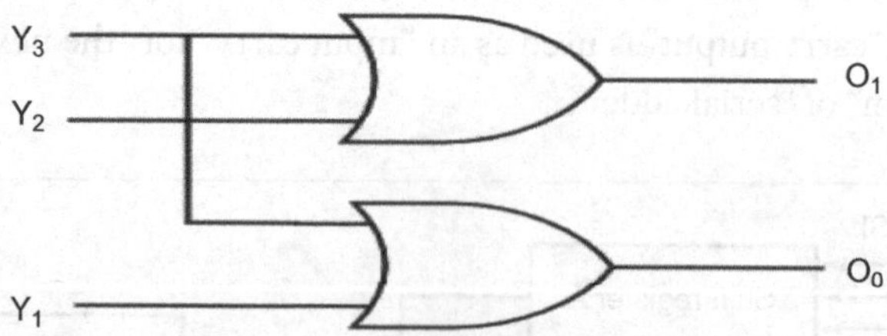

Priority Encoder: Scan for explanation

Fig. 3.14: Logic Circuit of 4:2 encoder

3.4.1 Priority Encoders

In normal encoder, if more than one inputs are high, then the encoder may produce "wrong output". It is the drawback of normal encoder. To overcome this issue, we assign priorities to the inputs, then the output will be according to the inputs which has higher priority. This encoder is called priority encoder. The "truth table" of "4:2 priority encoder" is shown in "Table 3.2". Here, the priority of Y3 is the highest and Y0 is the lowest. So, if Y3 is "1" the output is "11" irrespective of any other input. Sometimes a "valid" (V) signal is given at the output to validate the input set. For example if inputs are "0000" then the output is not valid. So, the valid signal "V" will be "0" in this case.

Table 3.2: "Truth Table" of "priority 4:2 encoder"

Y_3	Y_2	Y_1	Y_0	O_1	O_0
1	X	X	X	1	1
0	1	X	X	1	0
0	0	1	X	0	1
0	0	0	1	0	0

The Boolean expressions for O_1 and O_0 can be calculated using K-map-

$$O_1 = y_3 + y_2$$

$$O_0 = y_3 + y_2 . y_2$$

3.5 DECODERS

A "decoder" is a type of "combinational logic circuit" that decodes the "binary information on n lines" to "2n output lines". It performs the reverse operation of encoder circuit. It can be represented as n to 2n or n:2n decoder. A decoder can also be used with Enable (E) signal. When enable is HIGH, then it provides the outputs according to binary input information. The "block diagram" and "truth table" of 2:4 decoder is shown in fig. 3.15. It can be observed from the "truth table" and the "Boolean expressions" given in fig. 3.16 that the outputs of decoder are all possible minterms of the input set I1 and I2.

Inputs			Outputs			
E	I_1	I_0	Y_3	Y_2	Y_1	Y_0
0	X	X	0	0	0	0
1	0	0	0	0	0	1
1	0	1	0	0	1	0
1	1	0	0	1	0	0
1	1	1	1	0	0	0

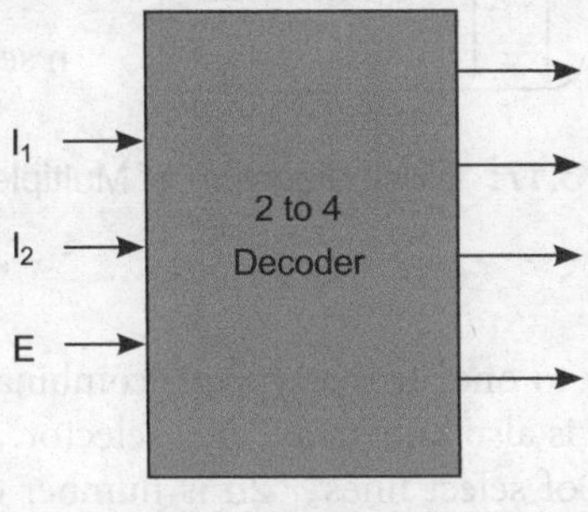

Fig. 3.15: "Block diagram" and ""truth table of 2:4 decoder

The "Boolean expressions" for the "outputs" and "logic circuit diagram" are given below-

$$Y_3 = E.I_1.I_0$$

$$Y_2 = E.I_1.\overline{I}_0$$

$$Y_1 = E.\overline{I}_1.I_0$$

$$Y_0 = E.\overline{I}_1.\overline{I}_0$$

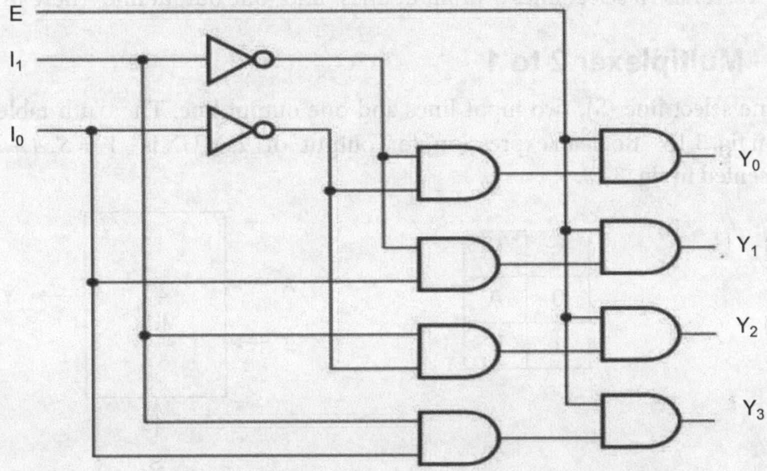

More about Decoder

Fig. 3.16: Implementation of 2:4 decoder using logic gates

3.5.1 Boolean logic implementation using Decoders

A "decoder" is also used to implement "Boolean functions". Since the outputs of "decoder" are representing all "minterms", therefore, it requires "OR" gate to generate the "sum of minterms" (SOP) form. The same is explained with the help of examples below.

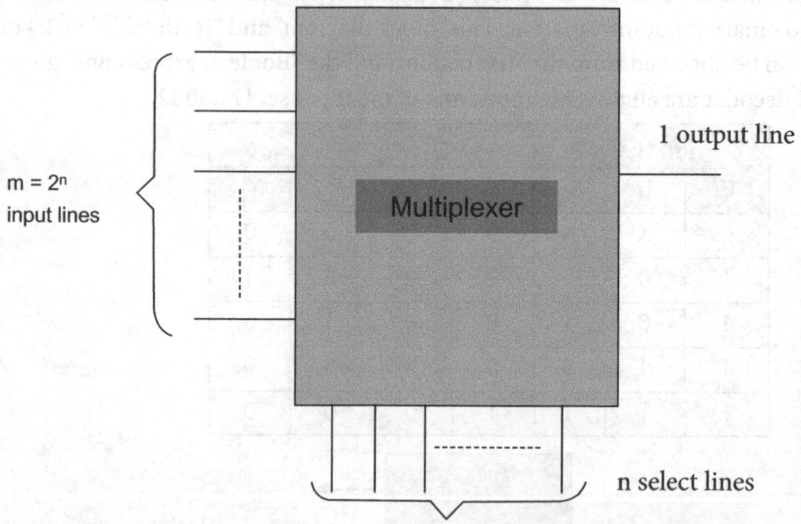

More about Multiplexer

Fig. 3.17: "Block diagram" of Multiplexer

3.6 MULTIPLEXERS

The term "Multiplex" means "many to one". It is a type of "combinational circuit" which has "multiple input lines" and "one output line". It is also known as "data selector" or "MUX" and represented as 2n:1 multiplexer, where, "n" is "number of select lines", "2n is number of input lines" and "1 output line". According to the values of select lines, the output gets value of one of the inputs. The types of multiplexers are "2:1 MUX", "4:1 MUX", "8:1 MUX" and so on. The general "block diagram" of a multiplexer is shown in fig. 3.18. It has "n select lines", "m input lines" and "one output line" (here m = 2n).

3.6.1 Multiplexer 2 to 1

It has one select line (S), two input lines and one output line. The truth table and block diagram are shown in fig. 3.18. "Boolean expression" for "output" of "2: MUX is"- $Y = \overline{S}.A + S.B$. The "logic diagram" is represented in fig. 3.19.

S	Y
0	A
1	B

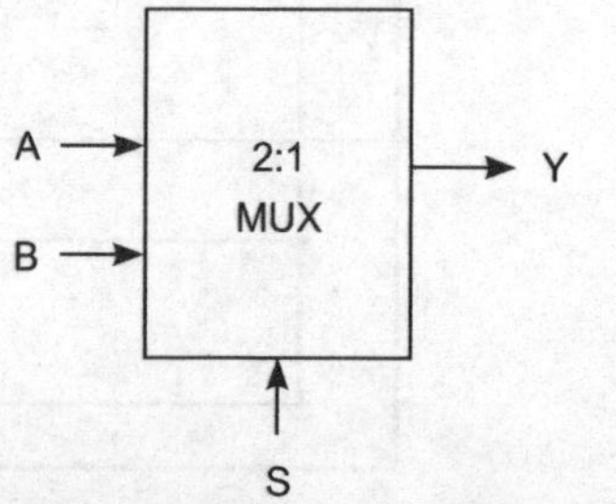

Fig. 3.18: Truth table and Block diagram of 2 to 1 MUX

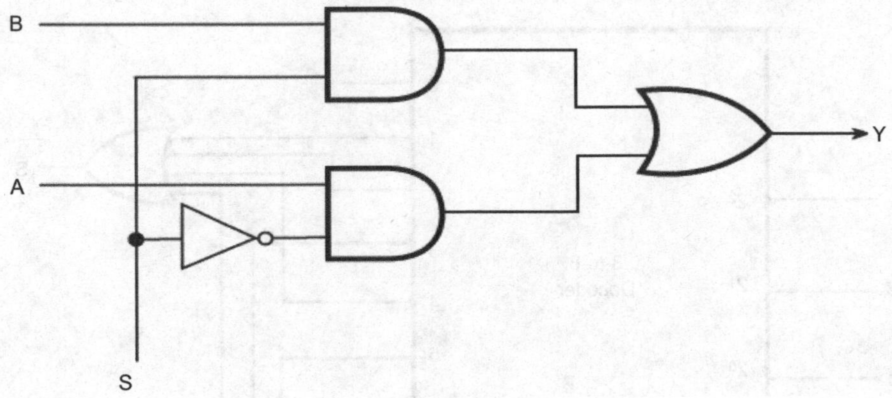

Fig. 3.19: Logic gate implementation of 2 to 1 MUX

Example 3.9

Implement following logic function using "3:8 decoder".

$$Y(A,B,C) = \Sigma(m0, m2, m4, m7)$$

Solution:

The decoder has "3-input lines" and "8-output lines". The "output Y" is the "OR" combination of minterms m_0, m_2, m_4, m_7.

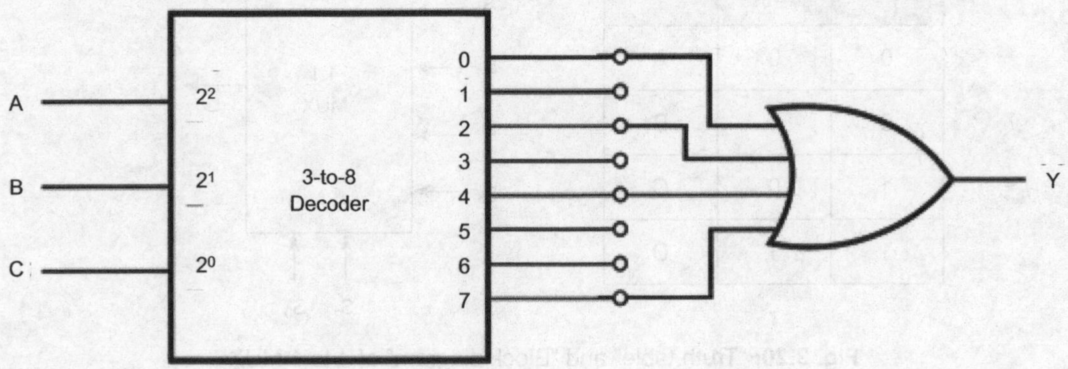

Example 3.10

Implement "full adder" using "3:8 decoder".

Solution:

From the "truth table" of "full adder", (see sec. 3.3.2)

Sum (S)= Σ(m1, m2, m4, m7) and Carry out (Co)= Σ(m3, m5, m6, m7). Therefore, it can be implemented as given below-

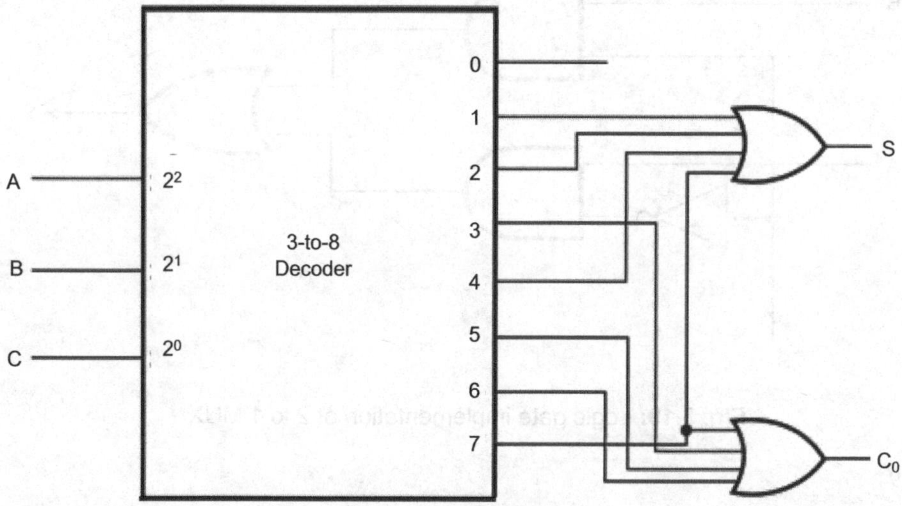

3.6.2 Multiplexer 4 to 1

It has "two select lines (S1 and S0)", "four input lines" and "one output line". The "truth table" and "block diagram" are shown in fig. 3.20.

S_1	S_0	Y
0	0	A
0	1	B
1	0	C
1	1	D

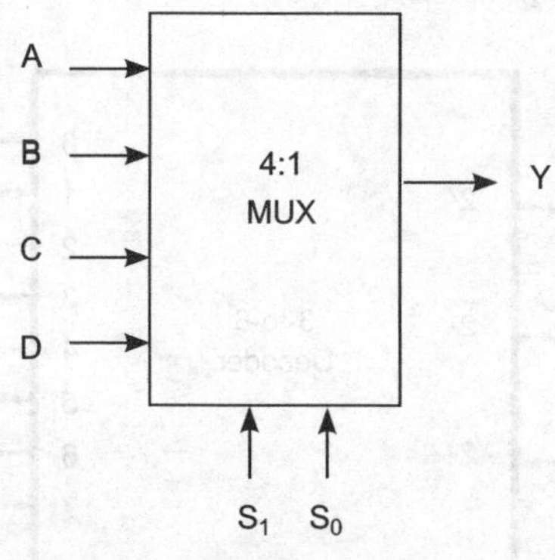

Fig. 3.20: "Truth table" and "Block diagram" of 4 to 1 MUX

The Boolean expression for Y is given as-

$$Y = \overline{S}_1.\overline{S}_1.A + \overline{S}_1.S_0.B + S_1.\overline{S}_0.C + S_1.S_0.D$$

Logic diagram for 4:1 MUX is represented in fig. 3.22.

3.6.3 Multiplexer 8 to 1

It has three select lines (S_2, S_1, and S_0), eight input lines and one output line. The "truth table" and "block diagram" are shown in fig. 3.22.

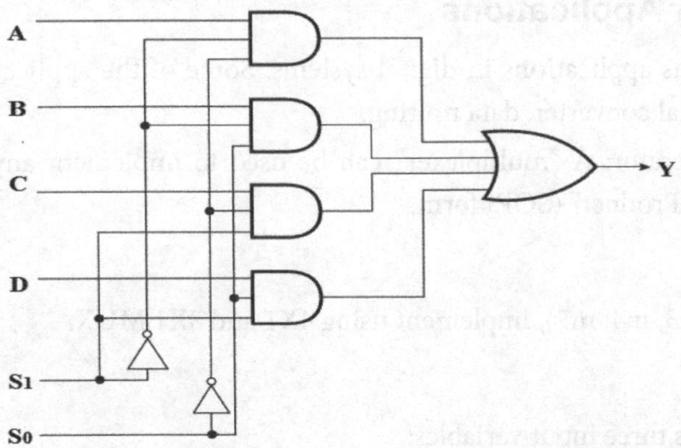

Fig. 3.21: Logic gate implementation of 4 to 1 MUXu

Select Lines ($S_2S_1S_0$)	Output (Y)
000	I_0
001	I_1
010	I_2
011	I_3
100	I_4
101	I_5
110	I_6
111	I_7

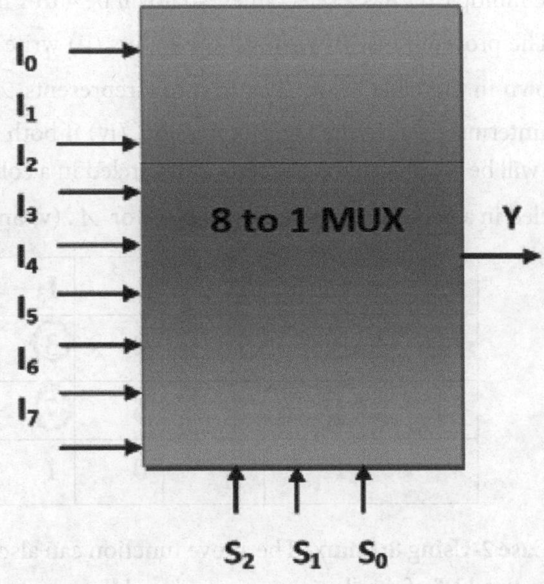

Fig. 3.22: "Truth table" and "block diagram" of 8 to 1 MUX

The "logic expression" for "output Y" of "8:1 MUX" is given below-

$$Y = I_0.\overline{S}_2.\overline{S}_1.\overline{S}_0 + I_1.\overline{S}_2.\overline{S}_1.S_0 + I_2.\overline{S}_2.S_1.\overline{S}_0 + I_3.\overline{S}_2.S_1.S_0 + I_4.S_2.\overline{S}_1.\overline{S}_0 + I_5.S_2.\overline{S}_1.S_0 + I_6.S_2.S_1.\overline{S}_0 + I_7.S_2.S_1.S_0$$

3.6.4 Boolean Logic Implementation using Multiplexers

A "multiplexer" can also be used to "implement" any "Boolean or logic function". If a logic expression has n+1 variables, then n variables can be used as "select lines" of the "multiplexer" and the remaining "single variable" can be used as "an input" along with its complement, "1 or Vdd" and "0 or ground". To understand the process, consider the example 3.11.

3.6.5 Multiplexer Applications

Multiplexers have various applications in digital systems. Some of the applications are logic function generator, Parallel to serial converter, data routing.

Logic function generator: A "multiplexer" can be used to implement any "logic function" from "truth table" in "Sum-of-Product" (SOP) form.

Example 3.11

Y (A,B,C)=Σ (m0, m1, m3, m4, m7), implement using 4X1 and 8X1 MUX.

Solution:

The above expression has three input variables;

Case 1-Using 4:1 mux- we will use two variables as select lines (B, C) and variable A as input line. As the multiplexer has 2 select lines, so it will be 4 to 1 multiplexer. Let say inputs are I_0, I_1, I_2, I_3.

The procedure is- (i) write all the inputs (ii) write all the minterms under the inputs in two rows as shown in the table below. The first row represents $\overline{A}$ and second row represents A. (iii) Circle all the minterms given in the logic expression. (iv) if both the minterms are not circled in a column, then input will be '0'; if both the minterms are circled in a column, then input will be '1'; if only one minterm is circled in a column, then input will be $\overline{A}$ or A. (v) and finally, write these inputs to the multiplexer.

	I_0	I_1	I_2	I_3
$\overline{A}$	(0)	(1)	2	(3)
A	(4)	5	6	(7)
	1	$\overline{A}$	0	1

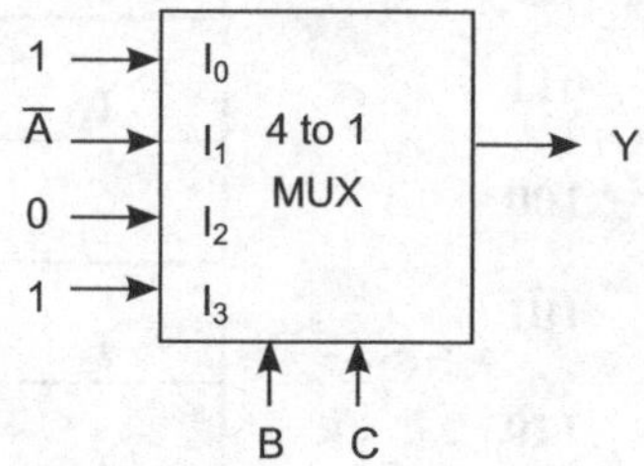

Case 2-Using 8:1 mux- The above function can also be "implemented" using "8:1 MUX". First draw the "truth table" of the "logic expression" and then write the inputs according the places from 0 to 7. This solution is shown on the next page.

Minterm	A	B	C	Y
0	0	0	0	1
1	0	0	1	1
2	0	1	0	0
3	0	1	1	1
4	1	0	0	1
5	1	0	1	0
6	1	1	0	0
7	1	1	1	1

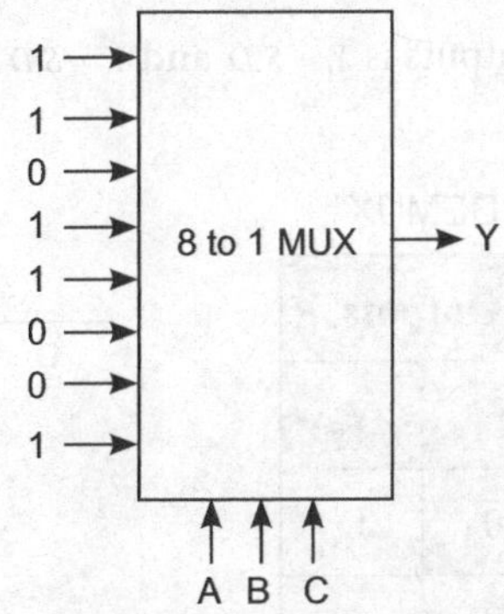

Parallel to serial converter: As the transmission of parallel data in digital system takes less time and data in parallel form can be converted into serial form using multiplexers.

Data routing: Multiplexers can be used to route data from multiple sources to single source.

3.7 DEMULTIPLEXERS

The term "Demultiplex" means "one to many". It is a type of "combinational circuit" which has "one input line" and "multiple output lines". It is also known as "DEMUX" and represented as "1:2n demultiplexer", where, "n" is "number of select lines", "2n" is "number of output lines" and "1- input line". Examples of "demultiplexers" are "1 to 2 demultiplexer", "1 to 4 demultiplexer", "1 to 8 demultiplexer", "1 to 16 demultiplexer" etc. The "block diagram" of demultiplexer is represented in fig. 3.24.

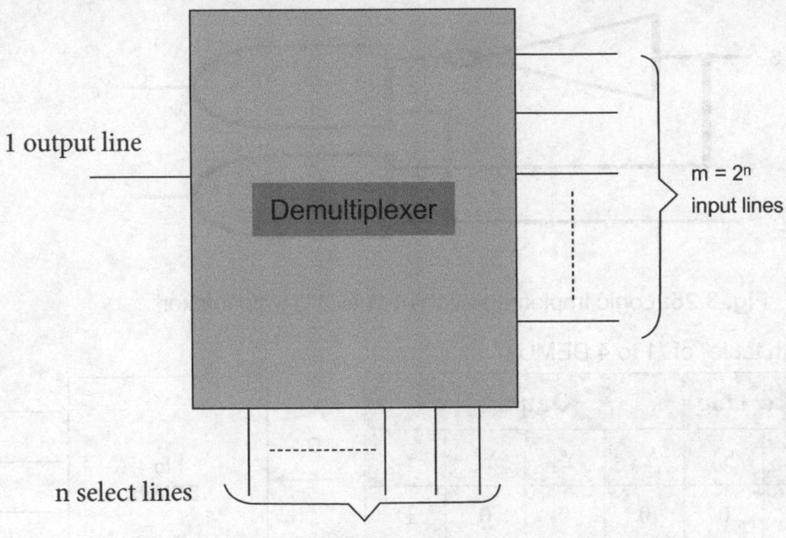

More about Demultiplexer

Fig. 3.23: "Block diagram" of Demultiplexer

3.7.1 Demultiplexer 1 to 2

1:2 demultiplexer has single input line, one select line (S) and two output lines. When S= '0', it will pass the data (D) to output line Y0 and if S= '1', it will pass the data (D) to output Y1. The truth table and block diagram are shown in Table 3.3 and Fig. 3.24 respectively.

The Boolean expression of the outputs is $Y_1 = S.D$ and $Y_0 = \overline{S}.D$. Logic gate implementation is given in fig. 3.26.

Table 3.3: "Truth table" of "1 to 2 DEMUX"

Input	Selecte Line	Outputs	
	S	Y_1	Y_0
D	0	0	D
D	1	D	0

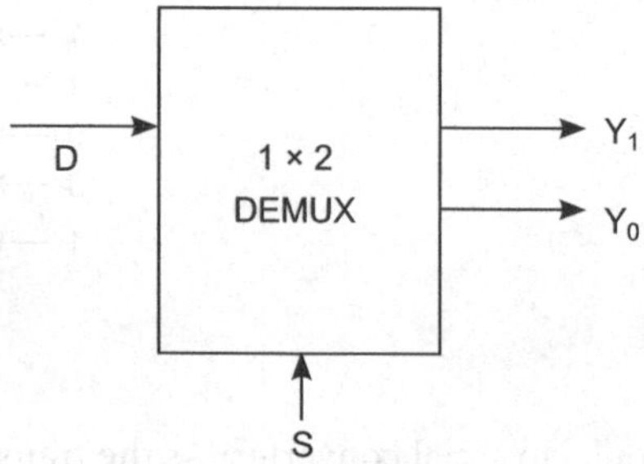

Fig. 3.24: "Block diagram" of 1 to 2 DEMUX

3.7.2 Demultiplexer 1 × 4

"1:4 demultiplexer" has "single input line", "two select lines (S1 and S0) " and "four output lines (Y3, Y2, Y1, Y0) ". When the selects lines are "00", the demultiplexer will provide the output at Y0; for select lines "01" it will provide the output at Y1; for select lines "10" it will provide the output at Y2 and for "11" it will generate the output at Y3. The "truth table" and " block diagram" are presented in "Table 3.4" and fig. 3.26 respectively.

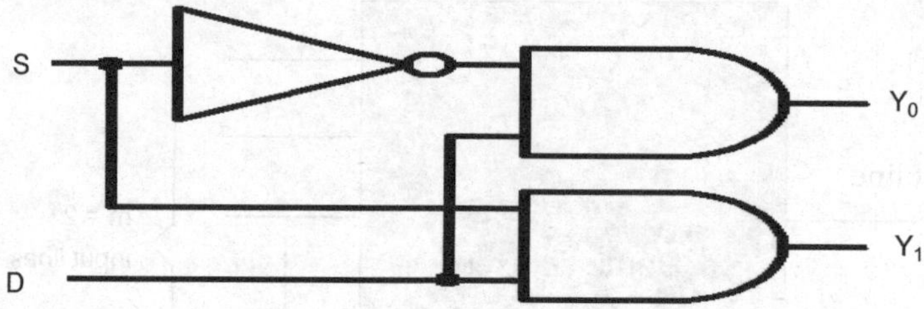

Fig. 3.25: Logic Implementation of "1 to 2 Demultiplexer"

Table 3.4: "Truth table" of "1 to 4 DEMUX"

Input	Selecte Line		Outputs			
	S_1	S_0	Y_3	Y_2	Y_1	Y_0
D	0	0	0	0	0	D
D	0	1	0	0	D	0
D	1	0	0	D	0	0
D	1	1	D	0	0	0

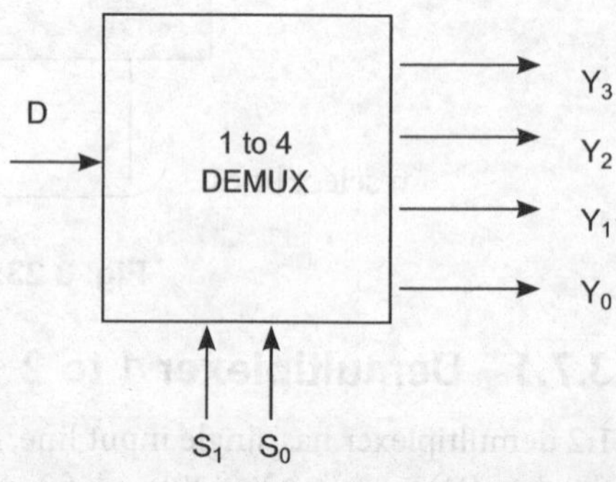

Fig. 3.26: Block diagram of "1:4 DEMUX"

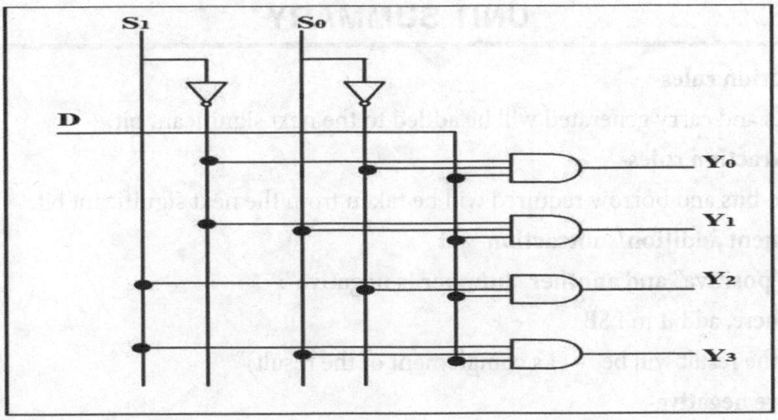

Fig. 3.27: "Logic Implementation" of "1 to 4 Demultiplexer"

Table 3.5: "Truth table" of "1 to 8 DEMUX"

Input	Selecte Line			Outputs							
	S_1	S_2	S_0	Y_7	Y_6	Y_5	Y_4	Y_3	Y_2	Y_1	Y_0
D	0	0	0	0	0	0	0	0	0	0	D
D	0	0	1	0	0	0	0	0	0	D	0
D	0	1	0	0	0	0	0	0	D	0	0
D	0	1	1	0	0	0	0	D	0	0	0
D	1	0	0	0	0	0	D	0	0	0	0
D	1	0	0	0	0	D	0	0	0	0	0
D	1	0	0	0	D	0	0	0	0	0	0
D	1	0	0	D	0	0	0	0	0	0	0

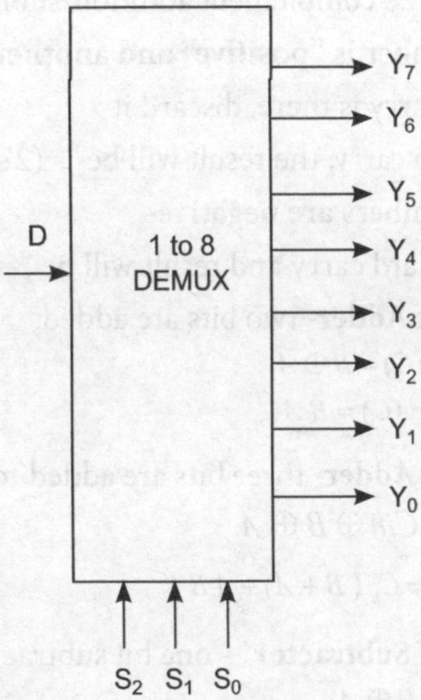

Fig. 3.24: "Block diagram" of "1 : 2 DEMUX"

$$Y_3 = S_1.S_0.D$$

$$Y_2 = S_1.\overline{S}_0.D$$

$$Y_2 = \overline{S}_1.\overline{S}_0.D$$

3.7.3 "Demultiplexer" 1 to 8

"1:8 demultiplexer" has "one input line", "three select lines (S2, S1 and S0) " and "eight output lines (Y7, Y6, Y5, Y4, Y3, Y2, Y1, Y0) ". The "truth table" and "block diagram" are shown in fig. 3.28.

The Boolean expression of the outputs is given below and logic gate implementation is given in fig. 3.28.

UNIT SUMMARY

- **Binary addition rules-**
  Add two bits and carry generated will be added to the next significant bit.
- **Binary subtraction rules-**
  Subtract two bits and borrow required will be taken from the next significant bit.
- **1's complement addition/subtraction**

One number is "positive" and another "number is negative":

- If Carry is there, add it to LSB
- If no carry, the result will be "- (1's complement of the result)"

Both numbers are negative-

- Add carry to LSB and result will be "- (1's complement of the result)"
  - 2's complement addition/subtraction

One number is "positive" and another "number is negative":

- If Carry is there, discard it
- If no carry, the result will be "- (2's complement of the result)"

Both numbers are negative-

- Discard carry and result will be "- (2's complement of the result)"
- **Half Adder**- two bits are added.
  $Sum(S) = B \oplus A$
  $Carry(C) = B.A$
- **Full Adder**- three bits are added together
  $S = Cin \oplus B \oplus A$
  $C_{out} = C_{in}(B + A) + A.B$
- **Half Subtractor** - one bit subtracted from another.
  $D = B \oplus A$
  $B_o = B.\overline{A}$
- Full Subtractor - two bits are subtracted from one bit.
  $Difference(D) = B_{in} \oplus B \oplus A$
  $Borrowout(B_o) = \overline{(A + B)}\, B_{in} + \overline{A}.B$
- **Encoder**
  An "encoder" is a type of "combinational circuit" that produces the "binary equivalent" to the "input". It has "2n input lines" and "n output lines".
- **Decoder**
  A "decoder" is a type of "combinational logic circuit that "decodes the binary information" on "n lines" to "2n output lines". It performs the "reverse operation" of an "encoder circuit". It can be represented as "n to 2n" or "n:2n decoder".

- **Multiplexer**
  The term "Multiplex" means "many to one". It is a type of "combinational circuit" which has "multiple input lines" and "one output line". It is also known as "data selector" or MUX and represented as "2n:1 multiplexer", where, "n is number of select lines", "2n is number of input lines" and "1 output line".
- **Demultiplexer**
  The term "Demultiplex" means "one to many". It is a type of "combinational circuit" which has "one input line" and "multiple output lines". It is also known as "DEMUX" and represented as "1:2n demultiplexer", where, "n" is "number of select lines", "2n" is "number of output lines" and "1-input line".

EXERCISES

Multiple Choice Questions

1. Binary addition of 110+010 is-
   (a) 1001 (b) 1000
   (c) 1100 (d) 1010
2. Binary subtraction of 1111-0101 is-
   (a) 1110 (b) 1101
   (c) 1010 (d) 1001
3. The result of $(12)_{10} + (8)_{10}$ is-
   (a) 10110 (b) 11101
   (c) 10111 (d) 10100
4. Which of the following represents Sum in half adder circuit, if A, B are the inputs-
   (a) AB + (BA)' (b) A'B+B'A
   (c) A'B'+BA (d) AB + (A+B)'
5. How many "data select lines" are required for "8X1" multiplexer?
   (a) 2 (b) 3
   (c) 1 (d) 4
6. A "full subtractor" can also be designed using-
   (a) "two half subtractors" and one "AND" gate
   (b) one half subtractor and two OR gates
   (c) "two half subtractors" and one "OR" gate
   (d) two half subtractors only
7. A "half adder" is a "combination" of-
   (a) XOR, "OR gate" (b) OR, "AND gate"
   (c) XOR, "AND gate" (d) only XOR
8. A full adder has "three inputs" and how many "outputs"?
   (a) one (b) three
   (c) four (d) two

9. $A.B+\overline{A}.\overline{B}$ "Two 4-bit $A.\overline{B}+\overline{A}.B$ binary $A.B+\overline{AB}$ numbers" $A.B+\overline{\overline{A}+B}$ ("1011 and 1111") are applied to a "4-bit parallel adder". The "carry input" is "1". Which is correct "sum" and "carry output"combination respectively?
   (a) "0000", "0" (b) "1111", "1"
   (c) "1011", "1" (d) "1100", "1"

10. A "full-adder" has a C_{in} = "0". Fing the "sum" (S) and the "carry" (C) if AB= 11?
   (a) S = 1, C = 1 (b) S = 1, C = 0
   (c) S = 0, C = 0 (d) S = 0, C = 1

11. Identify the following logic circuit-

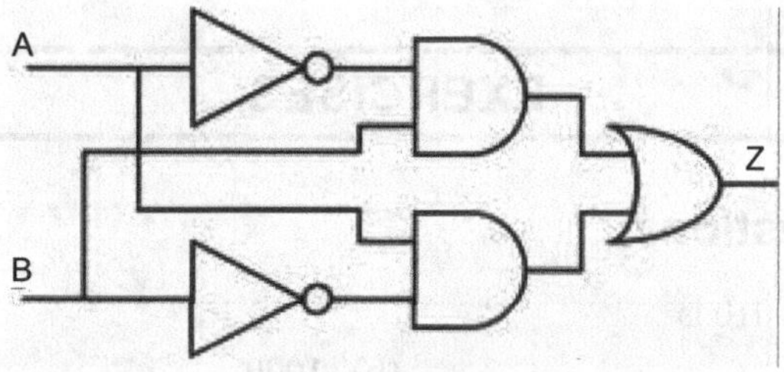

   (a) "OR gate" (b) "AND gate"
   (c) "XNOR gate" (d) "XOR gate"

12. The device that converts an inputs state into binary representation is known as-
   (a) encoder (b) decoder
   (c) MUX (d) DEMUX

13. For 8-input encoder, the number of possible combinations is-
   (a) 8 (b) 4
   (c) 24 (d) 28

14. If a multiplexer has 16 input lines, then number of select lines will be-
   (a) 3 (b) 2
   (c) 5 (d) 4

15. "Which of the following" circuit is known as "parallel" to "series converter"-
   (a) decoder (b) encoder
   (c) DEMUX (d) MUX

16. The number of select lines required for 1to 8 demultiplexer-
   (a) 4 (b) 2
   (c) 3 (d) 5

17. A 1 to 2 demultiplexer has ________ AND gates and ______ NOT gate-
   (a) 1, 1 (b) 2, 2
   (c) 3, 1 (d) 2, 1

18. "Which of the following circuit" can be used to implement the function :
   $F(A, B, C) = \Sigma(m_1, m_2, m_5, m_6)$
   (a) half adder (b) encoder
   (c) decoder (d) all

19. Identify the following circuit-

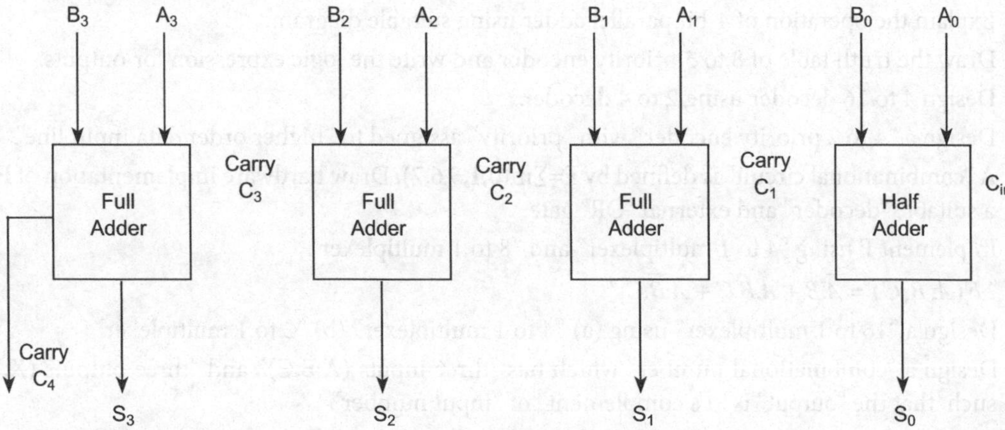

(a) parallel adder (b) serial adder
(c) full adder (d) encoder

20. A serial adder consists of __________ shift registers and ________ full adder-

(a) "2, 1" (b) "1, 2"
(c) "2, 2" (d) "1, 1"

ANSWERS															
1.	(b)	2.	(c)	3.	(d)	4.	(b)	5.	(b)	6.	(c)	7.	(c)	8.	(d)
9.	(c)	10.	(d)	11.	(d)	12.	(a)	13.	(d)	14.	(d)	15.	(d)	16.	(c)
17.	(d)	18.	(c)	19.	(a)	20.	(a)								

Short and Long Answer Type Questions

Category 1

1. Define combinational circuit and draw the block diagram.
2. Explain the rules of binary addition by taking an example.
3. Draw the truth table of half adder and write the logic expression of Sum and Carry.
4. How can we design Full adder using half adder? Draw the logic diagram.
5. Implement half adder using 2 to 4 decoder.
6. Implement half adder using multiplexer.
7. Implement the function F in the following circuit-

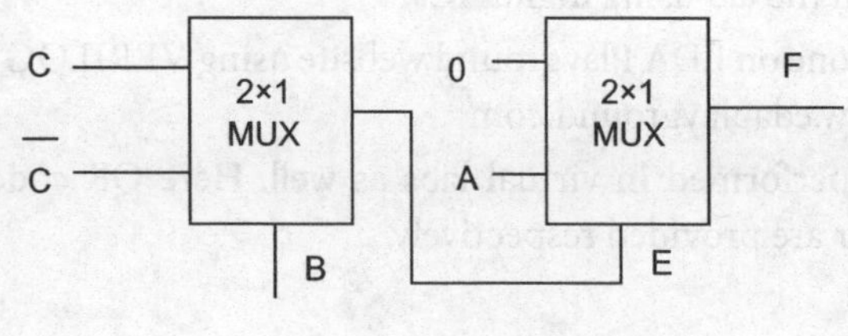

Category 2

1. Explain the operation of 4-bit parallel adder using suitable diagram.
2. Draw the truth table of 8 to 3 priority encoder and write the logic expression for outputs.
3. Design 4 to 16 decoder using 2 to 4 decoder.
4. Design a "4 to 2 priority encoder" with "priority" assigned to "higher order data input line".
5. A "combinational circuit" is defined by F=Σm(0,1,3,6,7). Draw hardware implementation of F with a suitable "decoder" and external "OR" gate.
6. Implement F using "4 to 1 multiplexer" and "8 to 1 multiplexer".
   " $F(A,B,C) = \overline{A}.B + A.B.\overline{C} + A.\overline{B}.C$ "
7. Design a "16 to 1 multiplexer" using (a) "4 to 1 multiplexer" (b) "2 to 1 multiplexer"
8. Design a "combinational number" which has "three inputs (A,B,C)" and "three outputs (X,Y,Z)", such that the "output" is "1's complement" of "input number".
9. Design a "combinational circuit" which a "three inputs" and "one output" such that the "output is HIGH" if the "input" is "greater than or equal to four".
10. Design a "logic diagram" of a "circuit for addition/subtraction" with a "control variable p". The circuit functions as "full adder" when "p=1" and as "full subtractor" with "p=0".
11. An "8 : 1 MUX" has "inputs A,B,C" connected to "select lines S_2, S_1, S_0" respectively. The data "inputs D_0 to D_7" are as follows: $D_1 = D_3 = 0$; $D_0 = D_2 = D_4 = D$; $D_5 = D_6 = D_7 = 1$. Determine the "minimized Boolean expression" that the MUX implements.

Numerical Problems

1. Solve using "2's complement"-
   (a) $(16)_{10} - (12)_{10}$ (b) $(20)_{10} - (24)_{10}$
2. (a) Perform the addition 101101 + 110011
   (b) Perform the subtraction 11011-1001
3. Solve using "1's complement"-
   (a) $(12)_{10} - (3)_{10}$ (b) $(11)_{10} - (18)_{10}$
4. Solve using "1's complement"-
   (a) $-(8)_{10} - (12)_{10}$ (b) $-(9)_{10} - (8)_{10}$
5. Solve using "2's complement"-
   (a) $-(10110)_2 - (1111)_2$ (b) $-(1010)_2 - (010)_2$

PRACTICAL

Experiments can be done in the lab using digital ICs.

-Experiments can be done on EDA Playground website using VERILOG language (Login is required using email ID).https://www.edaplayground.com

-Experiments can be performed in virtual labs as well. Here QR code for Virtual labs from IIT Bombay and IIT Kharagpur are provided respectively.

Virtual-Lab IIT Kharagpur

Virtual-Lab IIT Bombay

EXPERIMENT 3. "Implement Half Adder, Full Adder, Half Subtractor, Full subtractor using ICs".

Apparatus required-

Digital ICs- XOR -7486, AND- 7408, OR-7432, NOT-7404, NAND-7400. All ICs internal details are given in Experiment 1.

Diagram and Truth Table

Half adder-Please refer to section 3.3.1.

It requires one XOR gate (7486) for sum and one AND gate (7408) for Carry output. Sum = A XOR B; Carry = A AND B.

Alternatively "Half adder" can also be "implemented using NAND gate" (7400) only as shown in below diagram. In one IC there are 4 NAND gates, therefore two 7400 Ics are required to implement it.

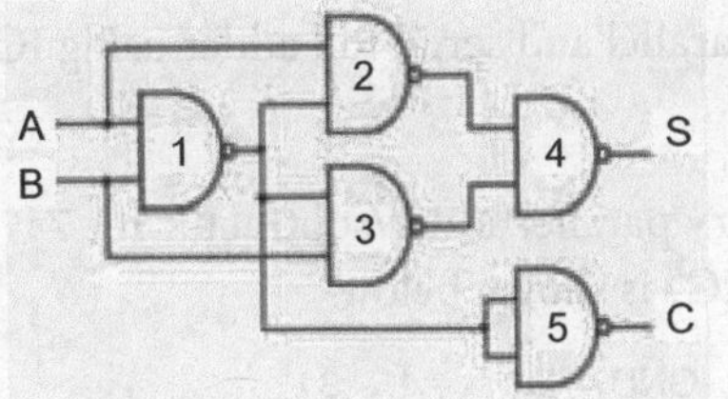

Full Adder – Please refer to section 3.3.2

It requires two half adders and one OR gate, fig. 3.7. Thus, it requires one XOR gate (7486) for sum and one AND gate (7408) and OR gate (7432) ICs for Carry output. Sum = (A XOR B) XOR Cin; Carry = (A AND B) OR ((A XOR B) AND Cin).

Alternatively Full Adder can also be implemented using NAND gate (7400) ICs only as shown in below diagram. Here 9 NAND gates or 3 NAND gate ICs 7400 are required to implement the Full Adder.

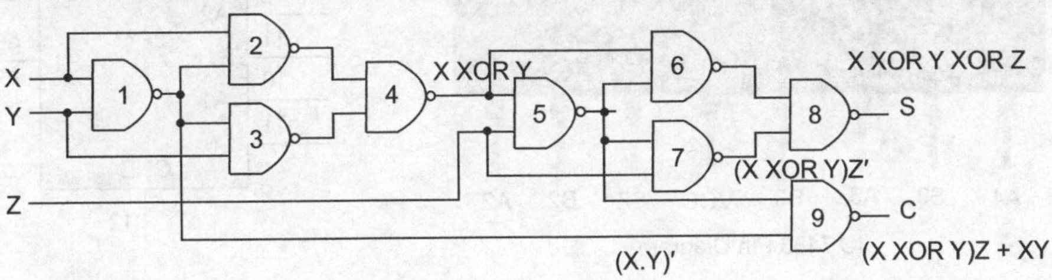

Half subtractor- Please refer to section 3.3.3.

It requires one XOR gate (7486) for Difference and one AND gate (7408) and one NOT gate (7404) ICs for Borrow output. Diff = A XOR B; Borrow = (NOT A) AND B, refer fig. 3.8.

Alternatively "Half subtractor" can also be "implemented using 5 NAND gates" (IC 7400) as shown here. In one IC there are 4 NAND gates, therefore two 7400 ICs are required to implement it.

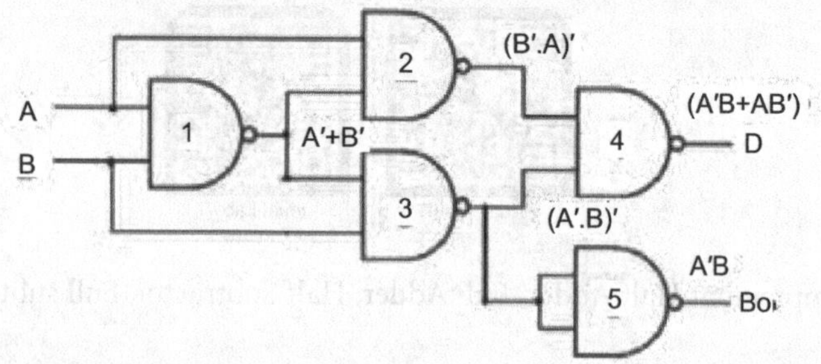

Full subtractor- Please refer to section 3.3.4.

It requires "two half subtractors" and "one OR gate", fig. 3.10. Thus, it requires one XOR gate (7486) for difference and two NOT gates (7404), two AND gates (7408) and one OR gate (7432) ICs for borrow (Bo).

Observations: Complete the truth table of each circuit by apply proper input pattern and observing the outputs and verify its correctness.

Conclusion: "Half adder", "full adder", "half subtractor" and "full subtractor" functions have been verified using digital ICs.

EXPERIMENT 4. "Implement parallel and serial full adder using ICs".

Apparatus required-

Digital ICs- 7483 is four bit binary parallel adder/subtractor IC. 7474- D flip flop IC, 7400-NAND gate ICs. The "pin diagram" of "7483 IC" is shown below-

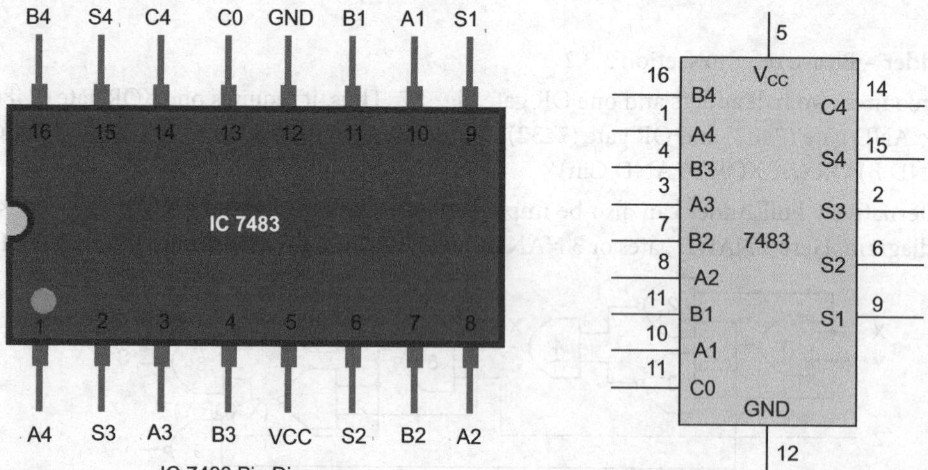

IC 7483 Pin Diagram

Procedure:

"Parallel Adder"- Please refer to section 3.3.5.

To perform four bit addition Pin 13 (Carry in) has to be grounded. Apply inputs A at pins 1,3,8,10 and B at pins 16,4,7,11 and observe the sum at pins 15,2,6,9 respectively and carry output at pin 14. How four bit adder internal circuit is given in fig. 3.12.

Serial Adder- Please refer to section 3.3.6.

Serial Full adder can be performed using the following diagram.

— If it is 4- bit full adder then first reset the D flip flop (so that Carry in = 0)
— Apply LSB of inputs A and B & Observe the output as LSB of sum
— Repeat the above step till MSB of inputs is reached.
— Carry out is the "output of D flip flop" and sum is the observations at "S output".

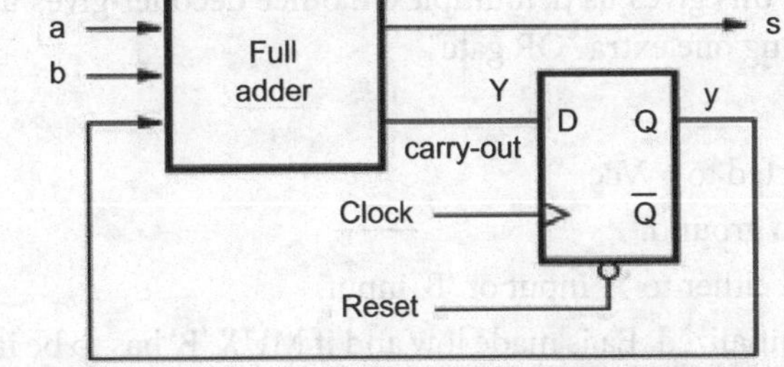

Here from the above diagram we can see that one- '1' bit full adder along with one D flip flop is required to implement serial adder.

Full adder circuit can be implemented as given in "EXPERIMENT 3" and to implement D flip flop (D -FF) we need IC 7474 which has two D-FF in it as shown in below pin and internal diagram.

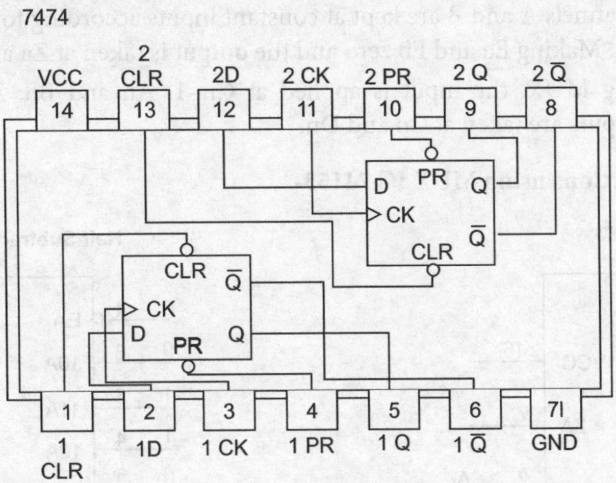

Observations: Complete the truth table by applying two or three sets of inputs and observe the outputs to verify its correctness.

Conclusion: Parallel adder and serial adder functions have been verified using digital ICs.

EXPERIMENT 5. "Design and development of Multiplexer and De-multiplexer using digital ICs. Implement basic circuits using these ICs".

Apparatus required-

Digital ICs- MUX IC-74153, Decoder IC- 74139, other basic gates ICs.

Multiplexer-

Details of the multiplexers are presented in sec 3.6. Digital IC for MUX is 74153 which has dual 4X1 MUX. Simple circuits like "Half Adder, Full Adder, Half Subtractor, Full subtractor" are implemented using 74153. The procedure for the IC connections is listed below-

Demultiplexer-

Demultiplexer is performed using 2X4 decoder IC 74139. IC 74139 is having dual channel decoders. It has enable input as well. If The input is applied at enable inputs and original inputs of decoders are used as selection lines then this gives us demultiplexer. Since decoder gives us minterms, so its easy to implement any function using one extra "OR gate".

Procedure MUX-

1. The Pin [16] is connected to + Vcc.
2. Pin [8] is connected to ground.
3. The inputs are applied either to 'A' input or 'B' input.
4. If MUX 'A' has to be initialized, Ea is made low and if MUX 'B' has to be initialized, Eb is made low.
5. Based on the selection lines one of the inputs will be selected at the output and thus the truth table is veried.
6. In case of half adder using MUX, sum and carry is obtained by applying a constant inputs at I_{0a}, I_{1a}, I_{3a} and I_{1b}, I_{2b} and I_{3b} and the corresponding values of select lines are changed as per table and the output is taken at Z0s as sum and Z0b as carry.
7. In this case, the channels A and B are kept at constant inputs according to the table and the input A and B are varied. Making Ea and Eb zero and the output is taken at Za and Zb.
8. In full adder using MUX, the input is applied at Cn–1, An and Bn. According to the table corresponding outputs are taken at Cn and Dn.

Implementation of functions using MUX IC 74153-

Half Adder Using 74153 –

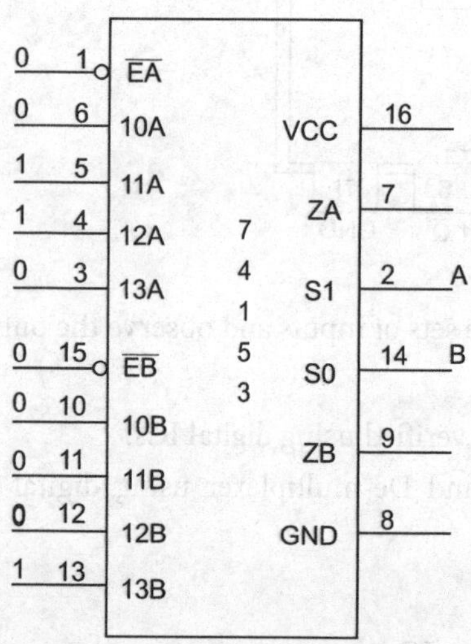

Half Subtractor: –

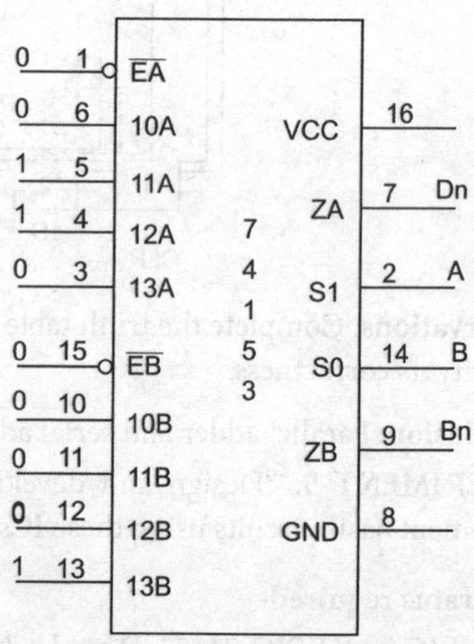

Full Adder Using 74153 –

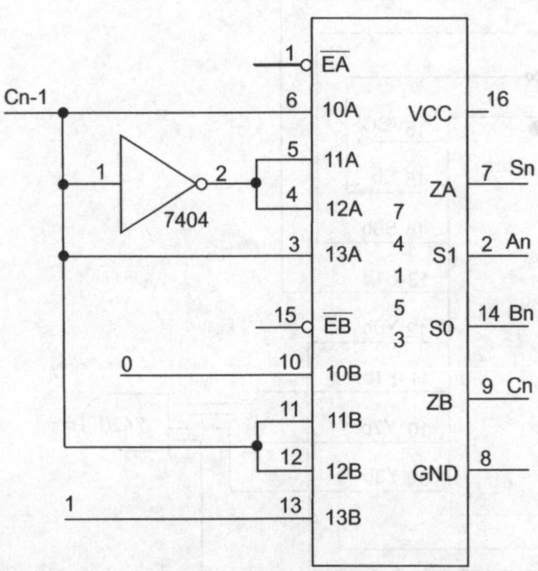

Full Subtractor Using 74153 –

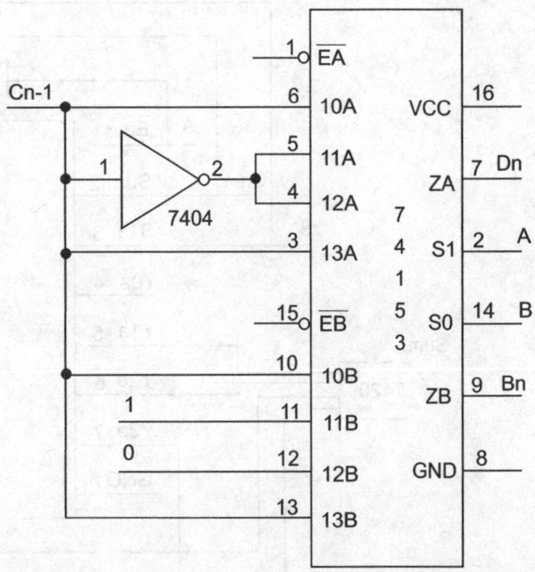

Procedure DEMUX uisng 74139 (decoder IC)-

1. The inputs are applied to either 'a' input or 'b' input
2. The demux is activiated by making Ea low and Eb low.
3. The truth table is verified.

Implementation of functions using DEMUX IC 74139-

Half Adder

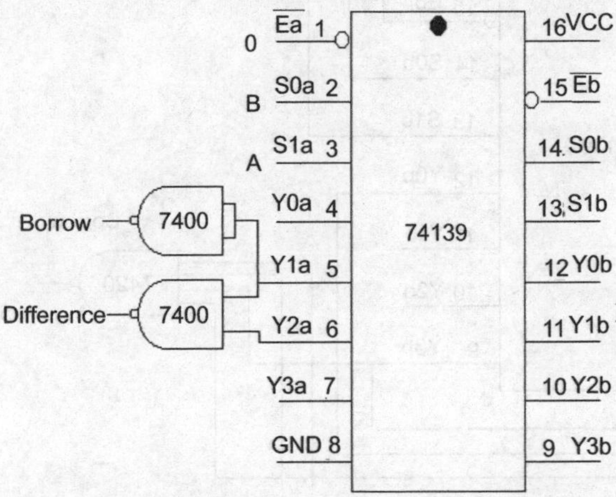

Half Subtractor

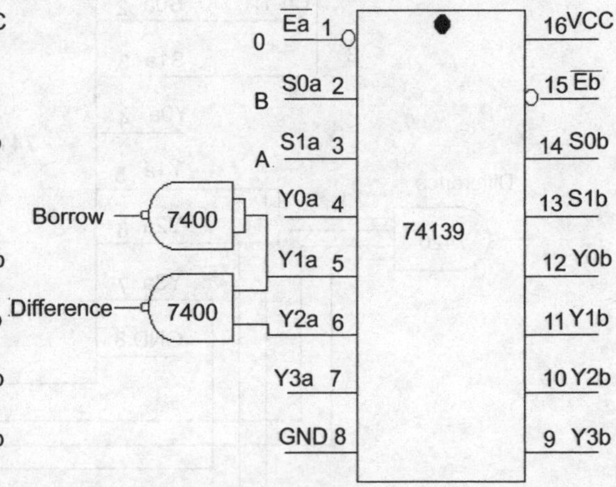

Full Adder-

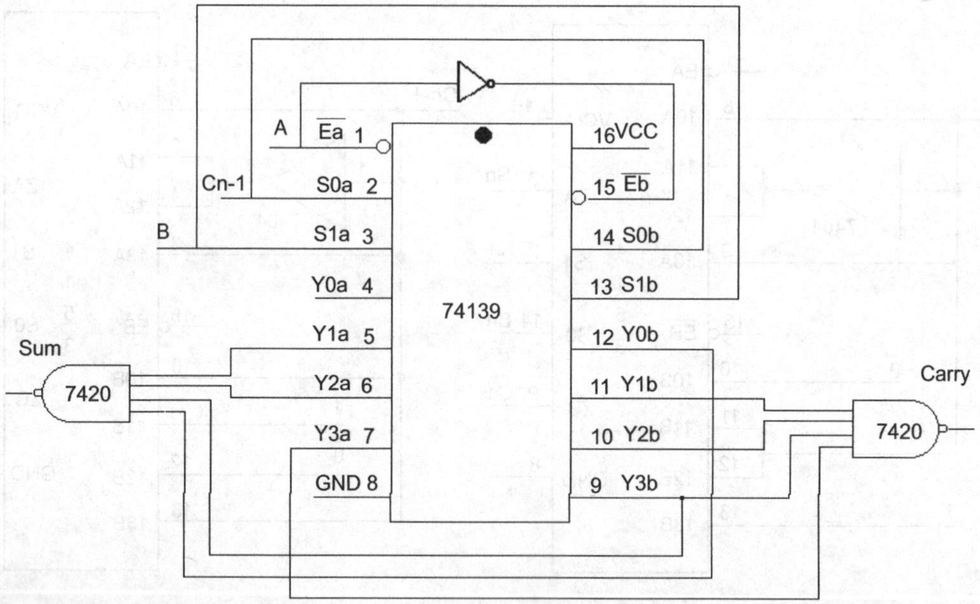

Full Subtractor-

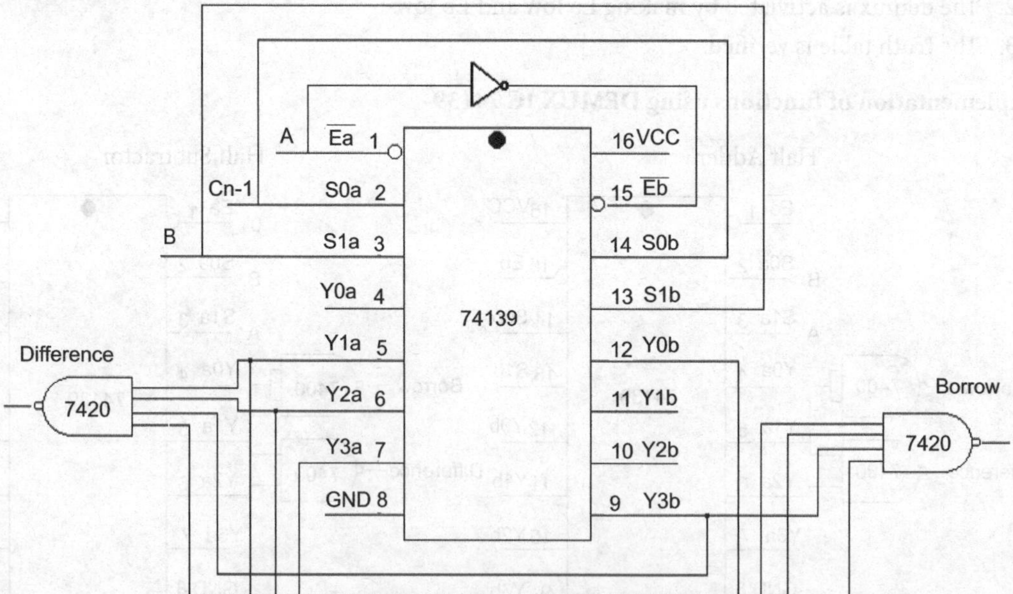

Observations: Complete the "truth table" of each circuit by applying proper input pattern and observing the outputs and verify its correctness refer sec 3.3 for "truth table" verification.

Conclusion: "Half adder, full adder, half subtractor and full subtractor" functions have been verified using digital ICs of MUX 74153 and DCODER (for DEMUX)-74139.

Verilog Experiments using online/offline software

- Important links- Yosys Open SYnthesis Suite :: Links (yosyshq.net)
- **Procedure-**
  1. Create your login at https://www.edaplayground.com/ and login.
  2. Choose SystemVerilog/Verilog from Language option given left side.
  3. Choose Tools & Simulators
     (a) Synthesis – "Yosys 0.9.0" -> tick on "Show diagram after run" Visit Yosys Open SYnthesis Suite :: About (yosyshq.net) for more help.
     (b) Simulation-"Icarus Verilog 0.9.7" -> tick on "Open EPWave after run"
  4. Left side is testbench space- type code here for testbench add "$dumpfile("dump.vcd"); $dumpvars(1);" after "initial begin". Save with testbench <module-name>.sv
  5. Right side is for RTL code- type "Verilog code for module", save with <module-name>.sv
  6. Remove the errors if any. Run and see the results.
- **Code-Half Adder using Verilog**

Data Flow modelling

```
module halfadder(a, b,
s, c);
input a;
input b;

output wire s;
output wire c;

assign {c,s}=a+b;

endmodule
```

Gate level Modeling

```
module halfadder(a, b,
s, c);
input a;
input b;
output s;
output c;

xor x1(s,a,b);
and a1(c,a,b);

endmodule
```

Testbench for Half adder

```
module halfadder_test();
reg a,b;
wire s,c;
halfadder h1(a,b,s,c);
 initial begin
     a=1'b1; b=1'b1;
#20 a=1'b1; b=1'b0;
#20 a=1'b0; b=1'b1;
#20 a=1'b0; b=1'b0;
end
 endmodule
```

Full Adder using Verilog

Data Flow Modeling

```
module full_adder
(carry,sum,x,y,z);
input x,y,z;
output
wire carry,sum;
assign {carry,sum
}=x +y+z;
endmodule
```

Gate level modeling

```
module full_adder(carry,s
um,A,B,Cin);
input A,B,Cin;
output carry,sum;
wire s1,c1,c2;
xor x1(s1,A,B);
and a1(c1,A,B);
xor x2(sum,s1,Cin);
and a2(c2,s1,Cin);
or o1(carry,c1,c2);
endmodule
```

Testbench Full Adder

```
module full_adder_test();
reg x,y,z;
wire carry,sum;
initial begin
 x=1'b0;  y=1'b0;  z=1'b0;
#20 x=1'b0;  y=1'b0;  z=1'b
1;
#20 x=1'b0;  y=1'b1;  z=1'b
0;
#20 x=1'b0;  y
=1'b1; z=1'b1;
end
endmodule
```

Any one among three modelling styles can be used to make the design. Testbench is used to see the waveform. For more details about Verilog you may visit the below webpage. Similarly other circuits can be created using Verilog.

Reference- https://www.asic-world.com/verilog/veritut.html

KNOW MORE

Combinational logic circuits are basic building blocks of all digital processors. Encoder are used to encode information so that it can be processed by digital processor while decoders are used to decode that information so that it can be received by the receiver in original format. For example mike encoded voice while before it goes to speaker it is decoded. Communication systems use multiplexers to accommodate more phone users in one channel while demultiplexers do the reverse at the receiving end.

REFERENCES AND SUGGESTED READINGS

1. M. Morris Mano and Michael Ciletti, "Digital Design", 5th edition, Pearson.
2. Digital Electronic Circuits - Course (nptel.ac.in)

Dynamic QR Code for Further Reading

-QR codes are embedded in the chapter

□□

4 Sequential Logic Circuits

UNIT SPECIFICS

Through this unit we have discussed the following aspects:

- Level-Triggered and Edge-Triggered Flip Flops (FFs)
- FF Timing Parameters,
- Master–Slave FFs
- FF Applications
- Counters
- Registers

The practical applications of the topics are discussed for generating further curiosity and creativity as well as improving problem solving capacity.

Besides giving a large number of multiple-choice questions as well as questions of short and long answer types marked in two categories following lower and higher order of Bloom's taxonomy, assignments through a number of numerical problems, a list of references and suggested readings are given in the unit so that one can go through them for practice. It is important to note that for getting more information on various topics of interest some QR codes have been provided in different sections which can be scanned for relevant supportive knowledge.

After the related practical, based on the content, there is a Know More section. This section has been carefully designed so that the supplementary information provided in this part becomes beneficial for the users of the book. This section mainly highlights the initial activity, examples of some interesting facts, analogy, history of the development of the subject focusing the salient observations and finding, timelines starting from the development of the concerned topics up to the recent time, applications of the subject matter for our day-to-day real life or/and industrial applications on variety of aspects, case study related to environmental, sustainability, social and ethical issues whichever applicable, and finally inquisitiveness and curiosity topics of the unit.

RATIONALE

This fundamental unit on Sequential Logic Circuits helps students to get a primary idea about the Level-Triggered and Edge-Triggered FFs, FFs – SR, J-K, T, D Flip Flop, J-K-Master Slave, FF Timing Parameters, R-S FF with Active LOW Inputs, R-S FF with Active HIGH Inputs,, Clocked R-S FF, J-K FF with PRESET and CLEAR Inputs, Master–Slave FFs, FF Applications, Triggering Counters – 4 bit Up

Down Counters, Asynchronous/ Ripple Counter, Decade Counter- Mod 3 Counter, Mod 7 Counter, Johnson Counter, Ring Counter, Registers – 4-bit Shift Register: Serial In Serial Out, Serial in Parallel Out, Parallel In Serial Out, Parallel In Parallel Out.

As the scope of digital electronics extend to domains such as healthcare, software, automation, digital banking and many more, various combination of logic gates are utilized to represent logic signal, known as sequential logic circuits. These are important part of digital electronics that are used in wide variety of applications.

PRE-REQUISITES

Unit 1, Unit II, Unit III.

UNIT OUTCOMES

List of outcomes of this unit is as follows:

U4-O1: Describe Level-Triggered and Edge-Triggered FFs

U4-O2: Learn the working of Master–Slave FF

U4-O3: Describe FF Applications

U4-O4: Learn the operation of Up and Down Counters

U4-O5: Designing the Series and Parallel Shift Registers

Unit-4 Outcomes	EXPECTED MAPPING WITH COURSE OUTCOMES (1 - Weak Correlation; 2 - Medium Correlation; 3 - Strong Correlation)					
	CO-1	CO-2	CO-3	CO-4	CO-5	CO-6
U4-O1	-	2	2	3	1	-
U4-O2	-	2	2	3	1	-
U4-O3	-	2	2	3	1	-
U4-O4	-	2	2	3	1	-
U4-O5	-	2	2	3	1	-

"Failure is the opportunity to learning again more intelligently."

-Henry Ford

4.1 INTRODUCTION

Input variables (X), computation circuit (logic gates), and output variables make up the sequential circuit as shown in Fig. 4.1. A sequential circuit generates an output based on both the primary input and previous output variables, as opposed to a combinational circuit which generates an output based only on the present input variables. Thus, memory components that can store binary information are present in sequential circuits.

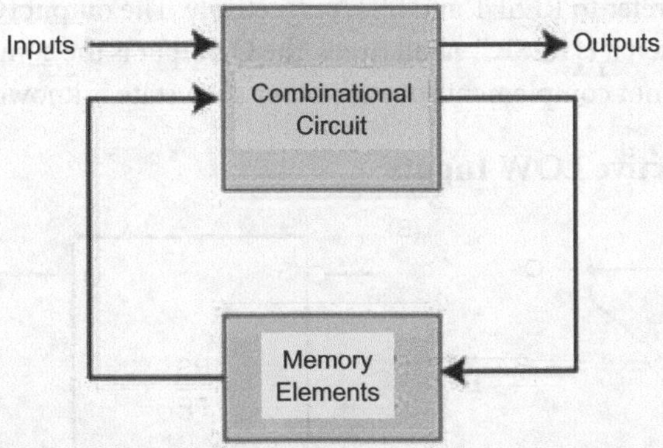

Fig. 4.1: Block Diagram for Sequential logic circuit.

4.1.1 Combinational Versus Sequential Logic Circuits

	Combinational Circuits	Sequential Circuits
1.	The output of the combinational circuit depend only on the present inputs.	The outputs of the sequential circuits depend on both present inputs and present state (previous output).
2.	The feedback path is not present in the combimantional circuit.	The feedback path is present in the sequential circuits.
3.	In combinational circuits, memory elements are not required.	In the sequential circuit, memory elements play an important role and require.
4.	The clock signal is not required for combinational circuits.	The clock signal is required for sequential circuits.
5.	The combinational circuit is simple to design.	It is not simple to design a sequential circuit.

4.2 FLIP FLOP

An example of a sequential circuit is a flip flop (FF) which normally takes its inputs and updates its results only at active edge of the clock. Latch is a level trigger sequential circuit. If in a FF triggering is done at active levels of clock, it is known as latch. Sometimes latch is also called level trigger FF.

In FFs, truth table is also called as characteristic table and Boolean equation of a FF is called as characteristic equation. FFs have state table and state diagrams to represent how the state transition is taking place depending on input combinations and current states. State excitation table or present state-next state table is used to design any sequential synchronous circuit using FF or converting one FF to work as other FF.

4.2.1 S-R Flip-flop

Fundamentally, an S-R latch that uses NAND gates and an extra enable input. It also is referred to as level-triggered SR-FF. In essence, if E = 1, this circuit will function as an S-R latch; but, if E = 0, the output remains unchanged.

R and S in this context refer to RESET and SET, respectively. The output (Q) changes from a 1 state when it is SET to a 0 state when it is RESET. At all times, the Q output is the complement of the Q' output. If for any input Q and Q'are not complements of each other such state is known as invalid state.

4.2.1.1 S-R latch with Active LOW Inputs

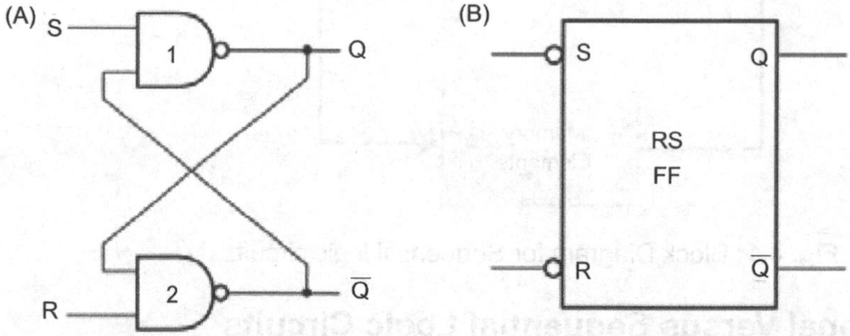

Scan for Flip Flops

Fig. 4.2: S-R latch (A) Circuit diagram and (B) Block Diagram with Active LOW Inputs

Cross-coupling connects the two NAND gates. In other words, one of NAND 2's inputs receives the output of 'NAND1' while the opposite is true of 'NAND2's' input. The 'S' and 'R' inputs are the final NAND1 and NAND2 inputs. Both Q and Q' outputs come from NANDs 1 and 2, Fig 4.2. Fig. 4.2 also shows the logic diagram and block diagram for Active LOW Inputs SR latch, Table 4.1(a) presents the Truth Table for S-R latch with Active LOW Inputs. Here S is used to SET and R is used to RESET the output Q. For active LOW S=0 is used to SET –Q=1, while R=0 is used to RESET- Q=0. Since S and R can't be active together therefore S=R=0 entry is prohibited for active LOW input S-R latch.

Table 4.1: Truth Table of S-R latch with Active LOW and Active HIGH

a) Active LOW

S	R	Q	Q'	Comments
0	0	X	X	Invalid
0	1	1	0	Set
1	0	0	1	Reset
1	1	Q	Q'	No Change

b) Active HIGH

S	R	Q	Q'	Comments
0	0	Q	Q'	No Change
0	1	0	1	Reset
1	0	1	0	Set
1	1	X	X	Invalid

4.2.1.2 S-R latch with Active High Inputs

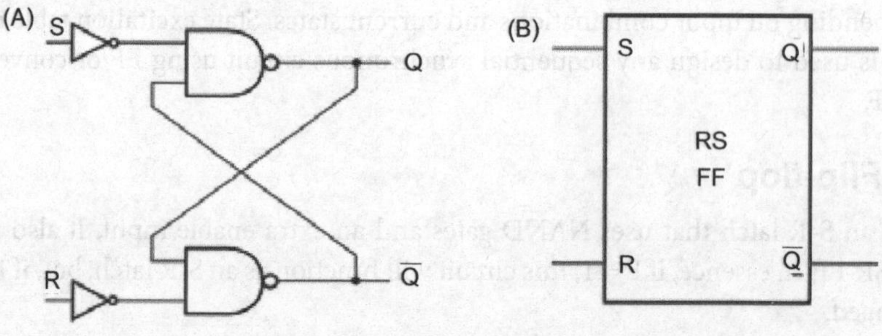

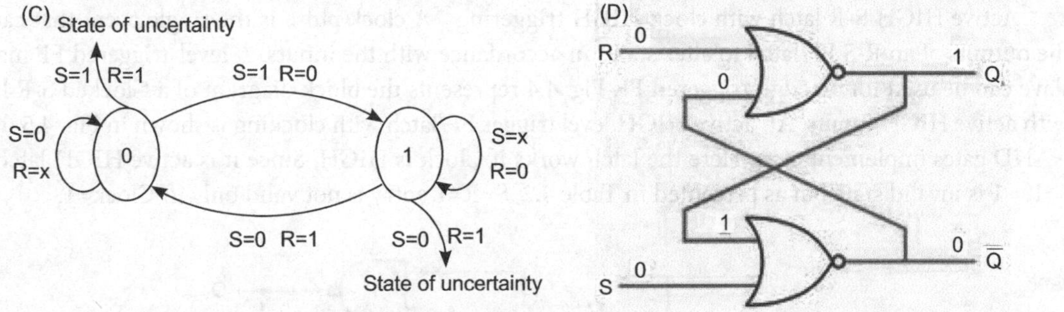

Fig. 4.3: S-R FF (A) Circuit diagram using NANDs (B) Block Diagram (C) state diagram (D) Circuit diagram using NORs with Active High Inputs

Since the S -SET and R -RESET inputs of an S-R latch can't be active simultaneously, therefore, entry R = S = 1 would be prohibited for active HIGH inputs. Table 4.1(b) presents the Truth Table for active HIGH S-R latch. Fig. 4.2 represents (A) NAND implementation logic diagram, (B) Block Diagram (C) State diagram and (D) NOR implementation logic diagram for active HIGH S-R latch. Here it can be noted the NOR based latch has interchanged S-R input terminals. S-R latch has two output states 0 and 1 as shown in state diagram here. If S=0 and state is Q= 0, it remains the same irrespective of R input. While S=1 makes a transition from state Q= 0 to 1 and the reverse is made by R= 1. Characteristic equation for S-R latch using k-map is given as –

Q = S+R'q or Q(t+1) = S+R'Q(t)

Where q or Q(t) is present state and Q or Q(t+1) is next state.

q \ SR	00	01	11	10
0	0	0	X	1
1	1	0	X	1

4.2.1.3 Clocked S-R latch or Level trigger FF

Table 4.2: Truth Table of Clocked S-R FF with active HIGH inputs

S	R	Clk	Q_{n+1}
0	0	0	Q_n
0	0	1	Q_n
0	1	0	Q_n
0	1	1	0
1	0	0	Q_n
1	0	1	1
1	1	0	Q_n
1	1	1	Invalid

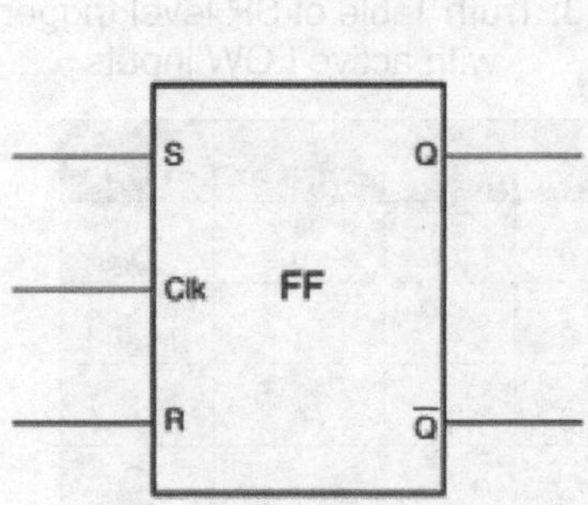

Fig. 4.4: Block Diagram of Clocked S-R FF with active HIGH inputs

Active HIGH S-R latch with clock HIGH triggering - A clock pulse is the single event that causes the outputs of an R-S FF/latch to alter states in accordance with the inputs. A level-triggered FF master slave can be used for an edge-triggered FF. Fig. 4.4 represents the block diagram of a Clocked S-R latch with active HIGH inputs. An active HIGH level trigger FF/latch with clocking is shown in Fig. 4.5 using NAND gates implementation. Here the latch works if Clock is HIGH. Since it is active HIGH latch, so S=R= 1 is invalid state but as presented in Table 4.2, S=R=1 entry is not valid only iff Clock=1.

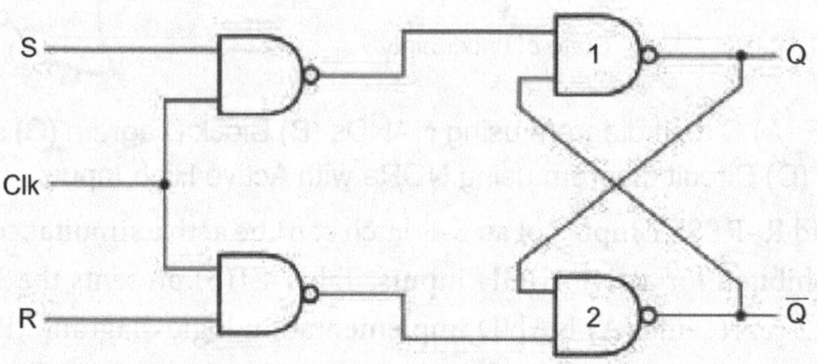

Fig. 4.5: Circuit diagram of SR level triggered FF/latch with active HIGH inputs.

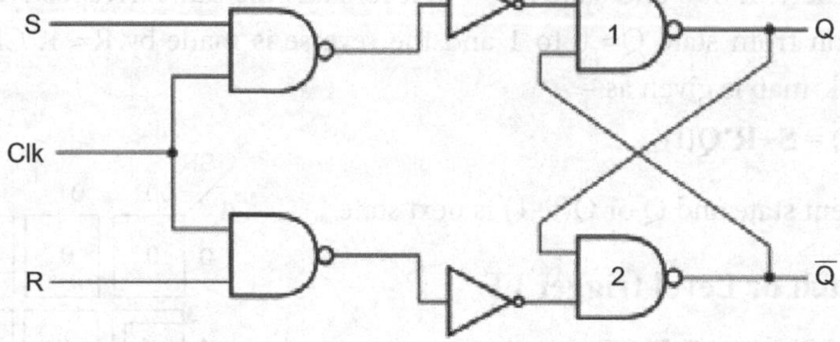

Fig. 4.6: Circuit diagram of SR level triggered FF/latch with active LOW inputs.

Table 4.3: Truth Table of SR level triggered latch with active LOW inputs

S	R	Clk	Q_{n+1}
0	0	0	Q_n
0	0	1	Invalid
0	1	0	Q_n
0	1	1	1
1	0	0	Q_n
1	0	1	0
1	1	0	Q_n
1	1	1	Q_n

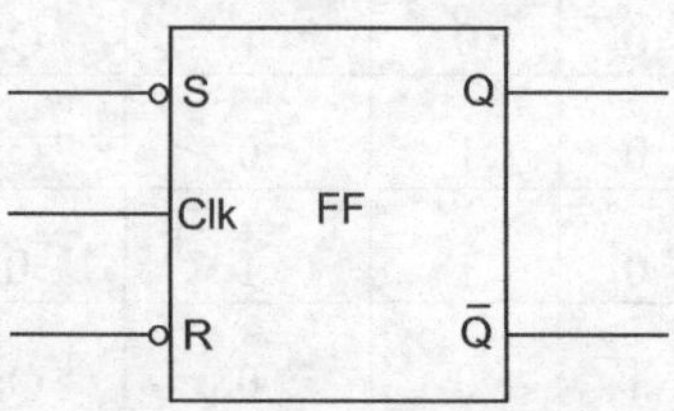

Fig. 4.7: Block Diagram of Clocked S-R FF with active LOW inputs

Active LOW S-R latch with clock HIGH triggering- The logic and block diagram are represented in Fig. 4.6 and Fig. 4.7 respectively. The truth table for this latch is presented in Table 4.3, it can be noted here that S=R=0 entry is not valid only iff Clock=1 for active low S-R latch.

4.2.3 Level-Triggered and Edge-Triggered FF

When the clock pulse level is active (HIGH or LOW), the output of a level-triggered FF/latch reacts to the information present at the inputs. In other words, any modifications made to the input while the clock is running (active level -HIGH or LOW) are mirrored in the output in accordance with its function table. In an edge-triggered FF, the output only reacts to data at the inputs on the clock signal's active edge (change from LOW to HIGH or –HIGH to LOW). The FF is referred to as positive edge activated in one case (change from- LOW to HIGH) and negative edge triggered in the other (change from- HIGH to LOW).

4.2.4 J-K FF -Positive clock edge triggered

In this subsection J-K FF active HIGH is discussed. Of all FF designs, the JK FF is the most popular and is regarded as a universal FF circuit. The JK FF is simply a gated SR FF designed such that it avoids the invalid output state that can happen when both inputs S and R are equal to logic level 1. A JK FF can accept four different input combinations to give outputs: logic 1, logic 0, no change, and toggle.

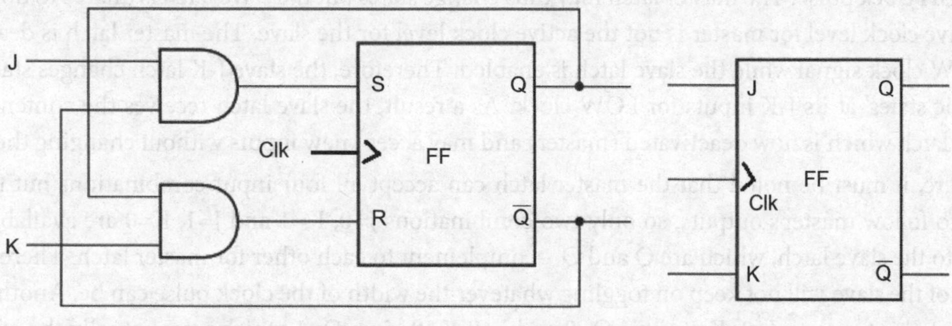

Fig. 4.8: Circuit diagram and Block Diagram of J-K FF

Block diagram and circuit diagram for positive edge triggered J-K FF is presented in Fig. 4.8. Truth Table for J-K FF is presented in Table 4.4. Here E –is enabling clock input. For positive level trigger latch E =1 while for edge triggered E is the active edge of the clock signal.

Race around condition: In J-K FF, it is very hard to achieve rising or falling edge. But if we make it level triggered as shown in Table 4.4, it is very hard to control the toggling of the output Q for a wider pulse width of clock signal and with J = K = 1 input set. This condition with active level of clock and J = K = 1 makes output to change continuously (toggle) for wider clock width. This condition is known as race around condition for J-K level trigger latch. This condition makes the J-K latch less reliable for use since the output is depending on how much it toggles depending on pulse width of the clock.

Table 4.4 : Truth Table of J-K FF

INPUTS			OUTPUTS		COMMENTS
E	J	K	Q	Q'	
	0	0	Q	Q'	NO CHANGE
	0	1	0	1	RESET
	1	0	1	0	SET
	1	1	Q'	Q	TOGGLE

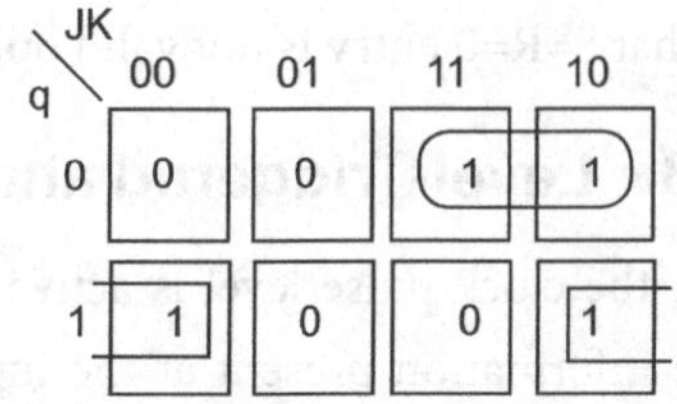

Characteristic equation for J-K- FF:

$Q = Jq' + K'q$

To avoid this race around condition in JK-latch, one solution is use of J-K edge triggered FF. But to generate active edge is really tough task. So this edge triggering can be achieved by Master Slave J-K FF. This will give solution to the race around condition and is actual JK- level triggered implementation separately for master and slave.

4.2.5 Master-Slave J-K FF

Two J-K latches are used to build a master-slave FF as presented in Fig. 4.9. The master latch is the first one while the slave latch is the second one. The clock for the master latch is HIGH level trigger while for slave latch it is LOW level trigger. The master latch is turned ON while the slave latch is OFF when there is a HIGH clock pulse. The master latch may thus change states but the slave latch is unable to do so since the active clock level for master is not the active clock level for the slave. The master latch is deactivated for LOW clock signal while the slave latch is enabled. Therefore, the slave J-K latch changes state as per the logic states' at its J-K inputs for LOW clock. As a result, the slave latch receives the contents of the master latch which is now deactivated (master) and may accept new inputs without changing the output.

Here, it must be noted that the master latch can accept all four input combinations but the slave needs to follow master's outputs, so only two combinations J=0, K=1 and J=1, K=0 are available as the inputs to the slave latch, which are Q and Q' –complement to each other for master latch. Therefore, the output of the slave will not keep on toggling whatever the width of the clock pulse can be. Another point to be noted is for slave J=0, K=1 gives Q=0 and J=1, K=0 gives Q=1 which were actually the outputs of the master latch at the falling edge of the clock. Since the outputs of second latch are the same as the first one (master latch), so the second latch is called as slave latch.To make a positive edge triggered J-K FF, one can apply Clk' to the master J-K latch and Clk to the slave J-K latch.

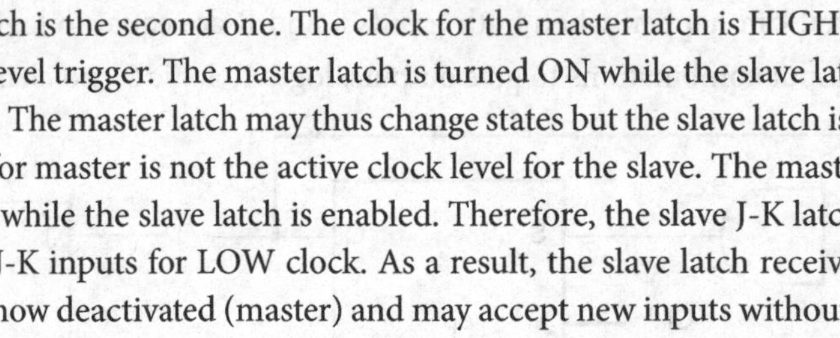

Fig. 4.9: Circuit diagram of J-K Master-Slave FF

4.2.6 Delay Flip Flop / D-FF

Delay FF, often known as D FF, is a basic gated S-R latch in which the S-R inputs are coupled to an inverter (NOT) gate. It just has one input D. After some time, the input data starts to emerge at the output. It is known as a delay FF because of the data delays from the input to output but does not change.

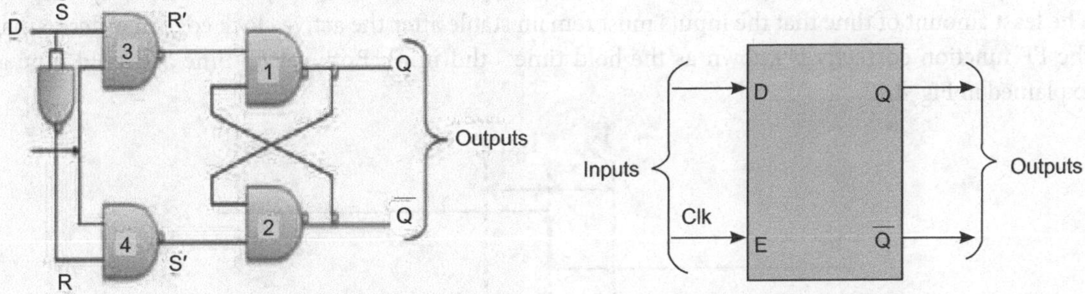

Fig. 4.10: Circuit diagram and Block Diagram for D FF

Due to the NOT gate, S and R will be the complements of one another as shown in Fig. 4.10. Therefore, neither S = R = 0 nor S = R = 1 will ever be an input condition, thus invalid state is avoided.

Table 4.5: Truth Table for D FF

INPUTS		OUTPUTS		COMMENTS
E	D	Q	Q'	
1	0	0	1	RESET
1	1	1	0	SET

Table 4.6: Truth Table of T FF

INPUTS		OUTPUTS		COMMENTS
E	D	Q	Q'	
1	0	Q	Q'	NO CHANGE
1	1	Q'	Q	TOGGLE

Thus the outputs will be just SET and RESET according D=1 or 0 respectively, as shown in Table 4.5. Characteristic equation for D-FF can be represented as – Q = D or Q(t+1) = D.

4.2.7 Toggle Flip Flop / T FF

Toggle FF is a JK FF with the J and K input terminals permanently tied together to make a single input. According to the Block Diagram shown in Fig. 4.11, it only has input symbolised by the letter T. The Block Diagram displays the symbol for a positive edge triggered T-FF. Since as the input is T=1 output Toggles as shown in Table 4.6, therefore, it is known as Toggle or T-FF. It has only two types of outputs either NO change in the output corresponding to T=0 or Toggle the output if T= 1 as shown in Table 4.6. Characteristic equation for T-FF can be written as XOR of T and q: – Q = Tq' + T'q.

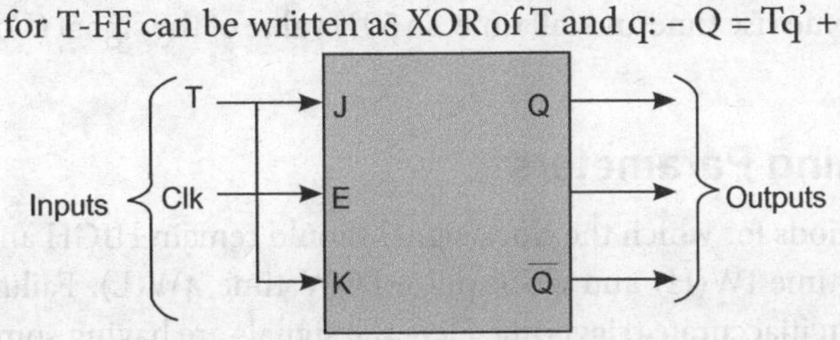

Fig. 4.11: Block Diagram of T (Toggle) FF

4.3 FF TIMING PARAMETERS

4.3.1 Set-Up Time and Hold Time

The set-up time is the least amount of time before the active clock transition for which the inputs must be steady (constant or no change) in order for the FF output to react correctly. It is denoted by tsu (min). The least amount of time that the inputs must remain stable after the active clock edge in order to make the FF function correctly is known as the hold time - thd(min). Both set up time and hold time are explained in Fig. 4.12.

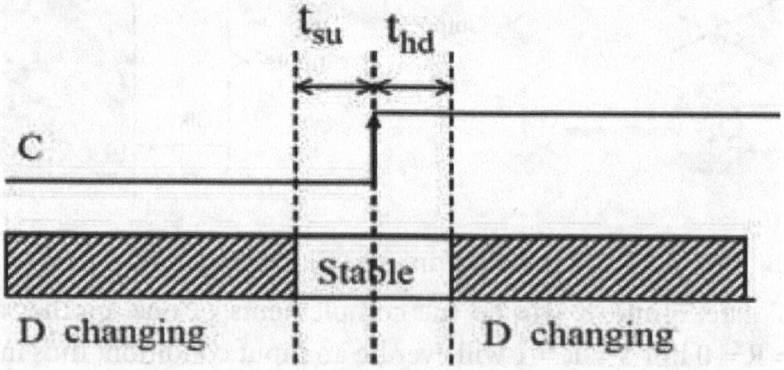

Fig. 4.12: Set-up time and hold time for a D- FF

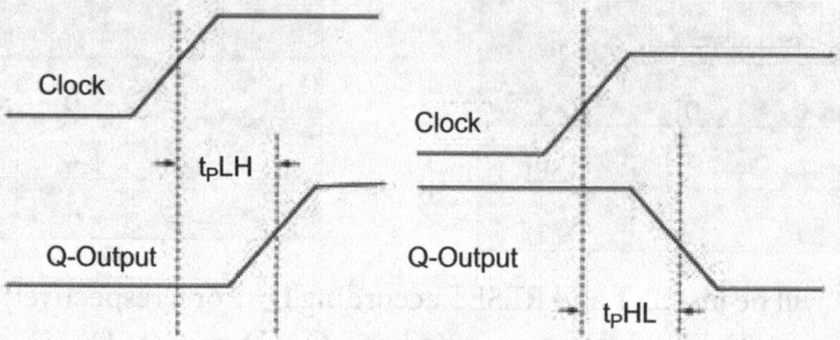

Fig. 4.13: Propagation delay

4.3.2 Propagation Delay

From the moment the signal is applied (active edge of the clock makes the input to activate the FF) until the output undergoes the desired change, there is always a delay known as the propagation delay. The propagation delays for both the HIGH-to-LOW (tpHL) and the LOW-to-HIGH (tpLH) output transitions are typically specified in the FF data sheet. Propagation delays are explained in Fig. 4.13. Here the signals are having some rise time and fall time, so 50% value of the signal (Vdd/2) is taken to calculate propagation delays.

4.3.3 Clock Timing Parameters

The minimum time periods for which the clock signal should remain HIGH and LOW are, respectively, the clock pulse HIGH time tW(H) and clock pulse LOW time, tW(L). Failure to comply with these standards might result in inaccurate triggering. Here the signals are having some rise time and fall time, so 50% value of the clock signal (Vdd/2) is taken to calculate Clock pulse times as shown in Fig. 4.14.

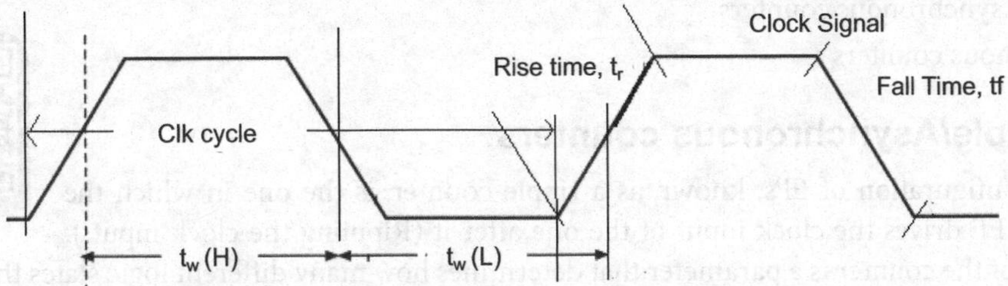

Fig. 4.14: Clock pulse HIGH and LOW, rise and fall times

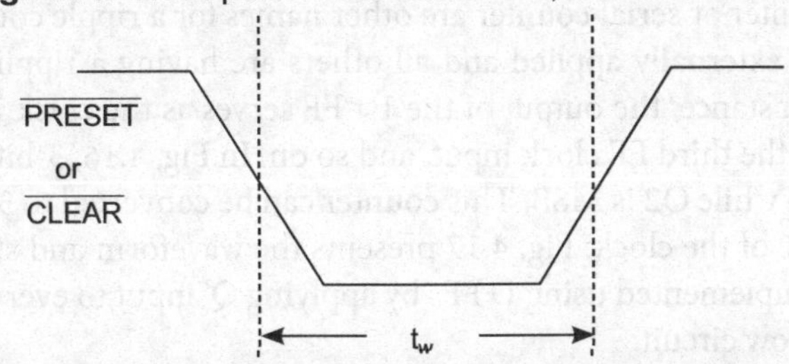

Fig. 4.15: Asynchronous input active pulse width

4.3.3.1 Clock Transition Times

Manufacturers give the maximum rise time (tr) and fall time (tf) that must pass for the output to react as intended. The FF may operate strangely or maybe not at all if these transition times are violated. These are represented in Fig. 4.14. There are different theories to calculate rise and fall times. Some calculate it from 0% to 100% (novice), experts calculate it from 10% to 90% and at low technology node it is calculated from 20% to 80% of Vdd.

4.3.3.2 Maximum Clock Frequency

The maximum frequency that may be applied to the clock input (frequency=1/Time period) is given by manufacturer. There is no assurance that the device will function properly and if frequency exceeds this amount.

4.3.3.3 Asynchronous Input Active Pulse Width

This is the least amount of time that the asynchronous input (PRESET or CLEAR) must remain active for the output to respond correctly, which is often LOW as presented in Fig. 4.15. PRESET and CLEAR are inputs for FFs which are active LOW. PRESENT is used to SET Q = 1 while CLEAR is used to set Q=0. Since both of these signals don't work with Clock synchronisation, therefore these are called Asynchronous Inputs.

4.4 DIGITAL COUNTERS

A counter is a sequential circuit. It is a type of digital circuit which is used to count pulses. The most widespread use of FFs is on counters. With a clock signal applied, it is a cluster of FFs. There are two types of counters -

(i) Ripple/Asynchronous counters

(ii) Synchronous counters

Digital Counters

4.4.1 Ripple/Asynchronous counters.

A cascaded configuration of FFs, known as a ripple counter, is the one in which the output of one FF drives the clock input of the one after it (Rippling the clock input). The modulus of the counter is a parameter that determines how many different logic states the cascaded arrangement passes through before repeating the sequence, determines how many FFs are present.

An asynchronous counter or serial counter are other names for a ripple counter in which the initial FF has clock input signal externally applied and all others are having a rippling clock obtained from previous FF's output. For instance, the output of the 1st FF serves as the clock input for the 2nd FF, the output of the 2nd FF feeds the third FF clock input, and so on. In Fig. 4.16, 3-bit up counter is presented using T-FF. Here Q0 is LSB while Q2 is MSB. This counter can be converted to 3-bit down counter using active edge as positive edge of the clock. Fig. 4.17 presents the waveform and state diagram a 3- bit up-counter. This can also be implemented using D FFs by applying Q' input to every D input of the same FF instead of T-FF used in below circuit.

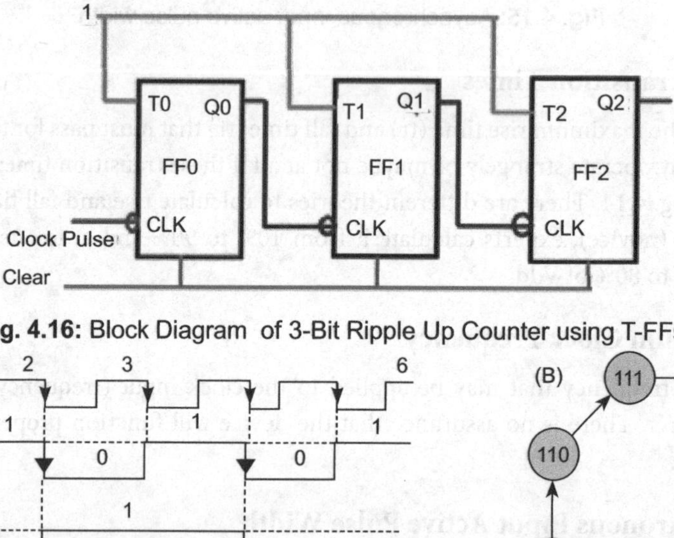

Fig. 4.16: Block Diagram of 3-Bit Ripple Up Counter using T-FFs

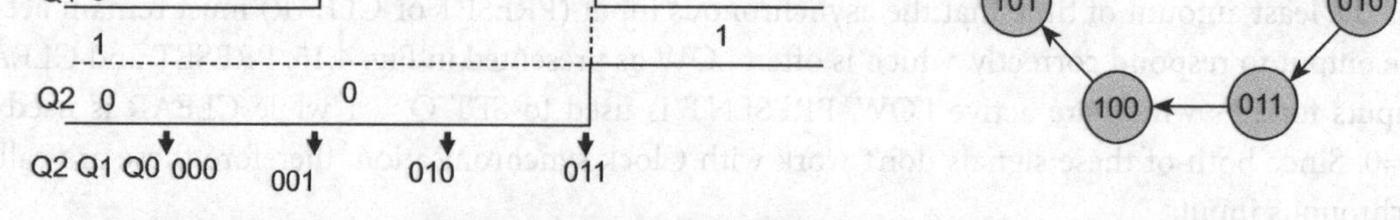

Fig. 4.17: (A) Wave diagram and (B) state diagram for 3-bit up counter

4.4.2 Synchronous Counter

All of the FFs of a synchronous counter are using same clock signal, sometimes referred to as a parallel counter. No matter how many FFs were used to build the counter, the delay in this situation is simply the propagation delay of one FF. In other words, the delay does not rely on how big the counter is.

4.4.2.1 Procedure for Designing Synchronous Sequential Circuits

The procedure for disigning synchronous sequential circuits can be summarized as follows:

1. From the word description and specifications of the desired operation, derive a state diagram for the circuit.
2. Reduce the number of states if necessary.
3. Assign binary values to the states.
4. Obtain the binary-coded state table.
5. Choose the type of flip-flops to be used.
6. Derive the simplified flip-flop input equations and output equations.
7. Draw the logic diagram.

   Synchromous counters can be designed using the above procedure. This will be disussed in the following subsections with example.

4.4.2.2 Modulus of a Counter

The number of various logic states that a counter passes through before returning to the beginning state and starting the count sequence again is known as the modulus (MOD number) of the counter. The modulus of an n-bit counter is the number of states it counts through without skipping any. As we can see, such counters have a modulus that is a power of two that is integral, i.e. 2, 4, 8, 16, and so on. To reach a modulus of less than 2n, these can be changed with the use of further combinational logic. Find the lowest integer m that is both equal to or larger than the desired modulus and also equal to the number of FFs needed to construct a counter with a specified modulus.

4.4.3 Modulo 3 (MOD-3) counter

Since 21<3< 22, therefore, minimum two FFs will be required to design MOD-3 counter and three FFs are required using one hot encoding. Assume that the counter counts three states 00, 01, and 10, the forth state 11 is not valid. The state diagram and state excitation table for MOD-3 counter is presented here in Fig. 4.18.

Synchronous counter- In this type counter, all FFs will use the same clock. The procedure to design any sequential circuit is presented in section 4.4.2.1. Table 4.7 presents the state excitation table for all FFs which will be used to create state excitation table for the sequential circuits.

Table 4.7 State excitation table for FFs

S-R FF Excitation Table

Q(t+1)	Q(t)	S	R
0	0	0	X
0	1	1	0
1	0	0	1
1	1	X	0

D-FF Excitation Table

Q(t+1)	Q(t)	D
0	0	0
0	1	1
1	0	0
1	1	1

JK-FF Excitation Table

Q(t+1)	Q(t)	J	K
0	0	0	X
0	1	1	X
1	0	X	1
1	1	X	0

T-FF Excitation Table

Q(t+1)	Q(t)	T
0	0	0
0	1	1
1	0	1
1	1	0

For MOD-3 counter, the state diagram, relation between states A, B, C excitation table and Boolean equations derived are in Fig. 4.18. Here we have assigned states A = 00, B= 01, C= 10. We have chosen minimum bits encoding to implement the counter, other way is one hot encoding in which as many bits as the number of states are required and only one bit is HIGH at a time. Here q1 is MSB and q0 is LSB for present states and Q1Q0 are for next states.

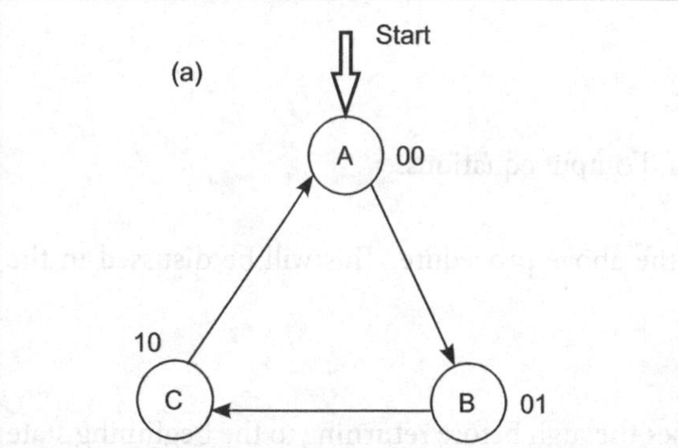

(b)

Present State	Next State
A	B
B	C
C	A

(c)

Present State (q_1q_0)	Next State (Q_1Q_0)	Input to FFs		
		Using D_1D_0	Using JK	
			J_1K_1	J_0K_0
00	01	01	0X	1X
01	10	10	1X	X1
10	00	00	X1	0X

(d) Boolean equations for FF inputs can be written using state excitation table-

$D_1 = q0$; $D0 = q1'q0'$;

$J1 = q0$; $K1 = 1$;

$J0 = q1'$; $K0 = 1$;

These expressions can be obtained using k-map.

Fig. 4.18: (a) State diagram, (b) Present state Next state relation and (c) state excitation table (d) Boolean equation for FF inputs for MOD-3 counter.

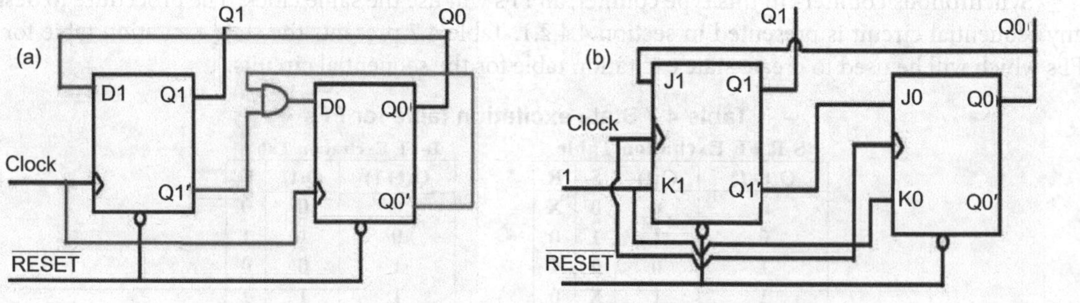

Fig. 4.19: Mod-3 counter (A) using D-FF, (B) using J-K FF .

4.4.3 Modulo 7 (MOD-7) counter

Since $2^2<7<2^3$, therefore, minimum three FFs will be required to design MOD-7 counter and 7 FFs are required for one hot encoding. Assume that the counter counts seven states from 000, to 110 states, the state 111 is not valid.

Asynchronous Design- This can be designed simply by using asynchronous up-counter. As soon as the counter reaches to the highest state (110 - in this case), the combination (110) can be applied to RESET or clear input which will bring it back to 000 state, as shown in Fig. 4.20.

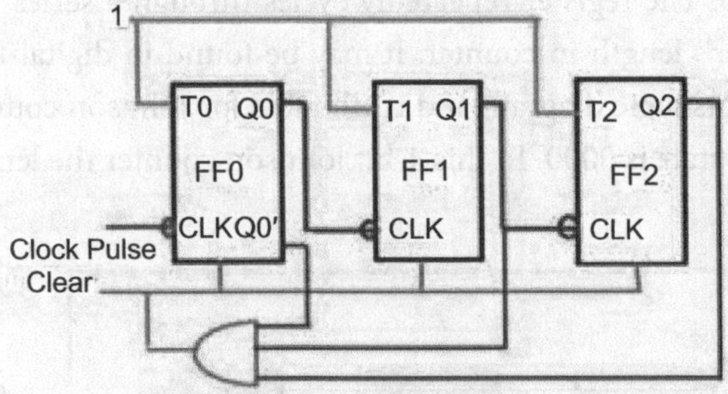

Fig. 4.20: Asynchronous MOD-7 counter

Note- If instead the states used are not continuous then we have to design either using synchronous counter using the given procedure or asynchronous counter with care to skip the intermediate invalid states. For example here instead of 111, assume 101 state is invalid. In this case asynchronous counter can be designed using PRESET and clear. As soon as counter reaches to 100, PRESET the Q_2Q_1 while CLEAR the Q0 to skip 101 and get 110 state in up-counter.

Locked state: if a sequential circuit once entered in invalid state remain in invalid states only, it is known as locked state. This has to be avoided while using don't cares for invalid states. It is explained in example 4.2.

4.4.5 Ring counter

A ring counter is a shift register with each FF's output coupled to its subsequent input, forming a ring. When n FFs are utilised, a single bit pattern is typically cycled to cause the state to repeat every n clock cycles. It is initially configured using PRESET and CLEAR inputs so that just one of its FFs is in state 1, while the rest are in state 0. Ring counter uses one hot encoding; therefore, the number of FFs will be equal to the number of states in the counter. Fig. 4.21 presents logic diagram and truth table for Ring counter. Here FF0 is PRESET to 1 at the starting and all others are CLEARed to 0 using ORI input signal, so starting state is 1000.

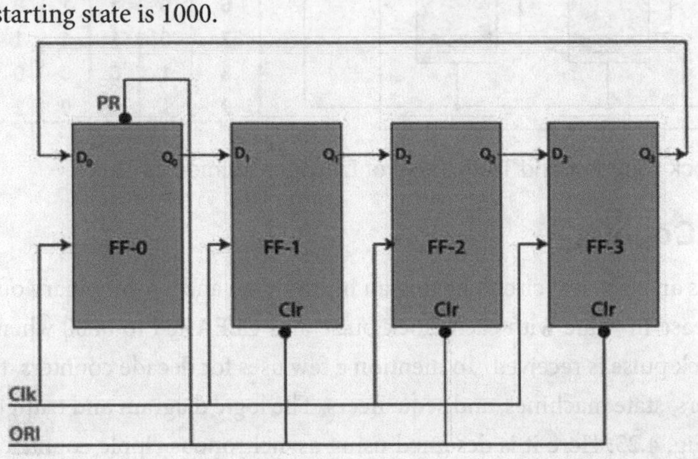

State	Q_0	Q_1	Q_2	Q_3
A	1	0	0	0
B	0	1	0	0
C	0	0	1	0
D	0	0	0	1

Fig. 4.21: Block Diagram and Truth Table of 4-Bit Ring Counter

4.4.6 Johnson Counter

A modified ring counter is known as a Johnson counter inverts the output of the last step before feeding it back into the initial flop. The register repeatedly cycles through a series of bit patterns that is twice as long as the shift register's length in counter. It may be found in digital-to-analog converters rather frequently. Fig. 4.22 presents logic diagram and truth table for Johnson counter. At starting every FF is CLEARed to 0, so starting state is 0000. In this 4-bit Johnson counter the length is 4 and different states are 8.

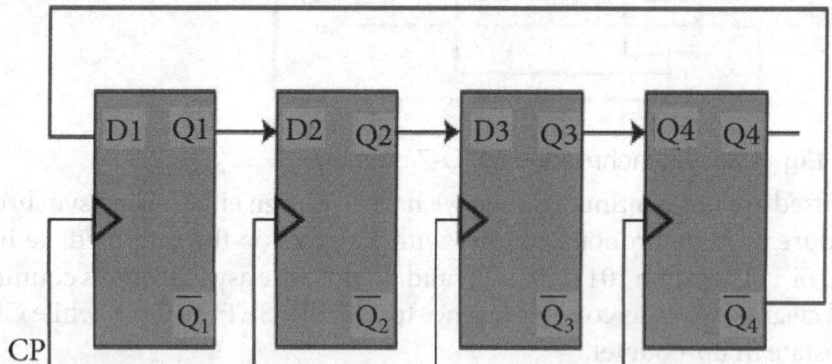

State	Q1	Q2	Q3	*Q4*
A	0	0	0	0
B	1	0	0	0
C	1	1	0	0
D	1	1	1	0
E	1	1	1	1
F	0	1	1	1
G	0	0	1	1
H	0	0	0	1
I	0	0	0	0

Fig. 4.22: Block Diagram and Truth Table of 4-Bit Johnson Counter

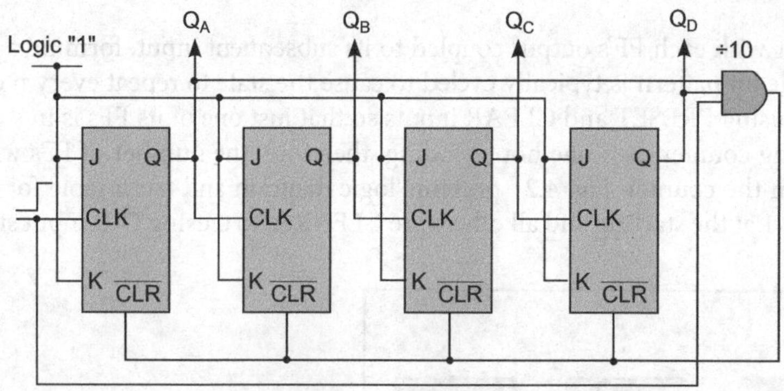

State	QA	QB	QC	QD
0	0	0	0	0
1	0	0	0	1
2	0	0	1	0
3	0	0	1	1
4	0	1	0	0
5	0	1	0	1
6	0	1	1	0
7	0	1	1	1
8	1	0	0	0
9	1	0	0	1

Fig. 4.23: Block Diagram and Truth Table of Decade Counter

4.4.7 Decade (mod-10) Counter

The fundamental decade counter is an electrical circuit having an input signal and a 4-bit binary output (called a clock). The outputs increase in value with each clock pulse and CLEARed to 0000 when the output reaches 1001 and a new clock pulse is received. To mention a few uses for decade counters, there are clock circuits, frequency dividers, state machines, and sequencers. The logic diagram and truth table for decade counter are shown in Fig. 4.23. Here it is designed using asynchronous ripple counter. The same can also be designed using synchronous procedure listed in section 4.2.2.1.

Example 4.1

Complete timing diagram for the circuit shown here.

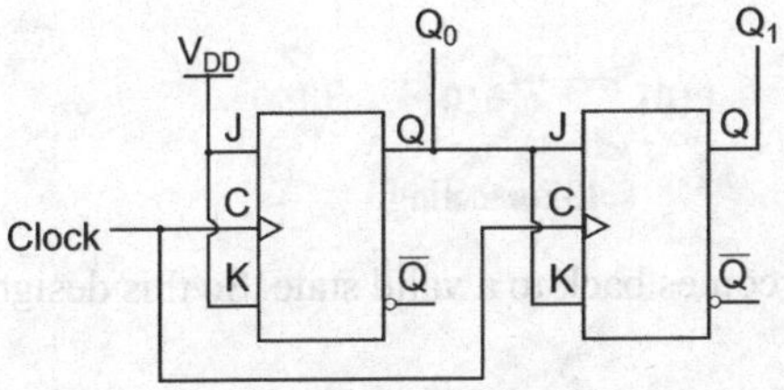

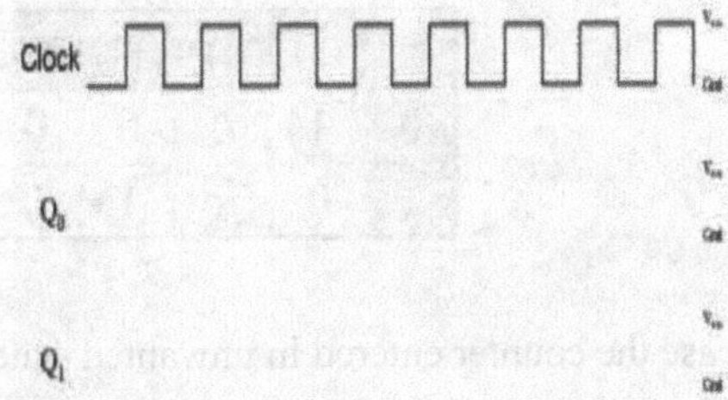

Solution:

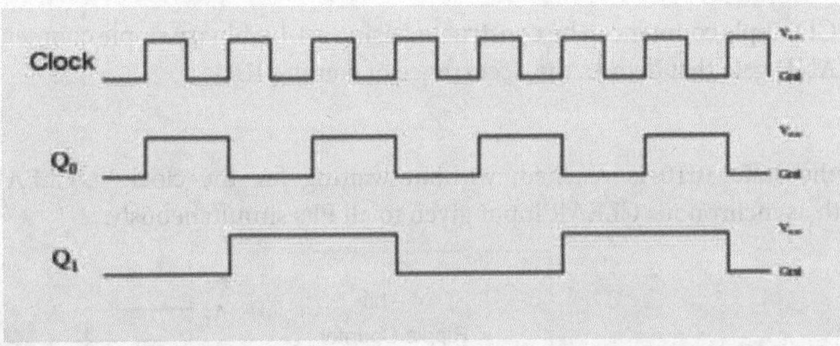

Example 4.2

Design a counter with T FFs that goes through the following binary repeated sequene: 0, 1, 3, 7, 4. Show that when binary states 010 and 101 are considered as don't care conditions, the counter may not operate properly. Find a way to correct the design.

Solution:

Present State			Next State			Flip-Flop Inputs		
A	B	C	A	B	C	T_A	T_B	T_C
0	0	0	0	0	1	0	0	1
0	0	1	0	1	1	0	1	0
0	1	0	X	X	X	X	X	X
0	1	1	1	1	1	1	0	0
1	0	0	0	0	0	1	0	0
1	0	1	X	X	X	X	X	X
1	1	0	1	0	0	0	1	0
1	1	1	1	1	0	0	0	1

$T_A = A \oplus B; T_B = B \oplus C$

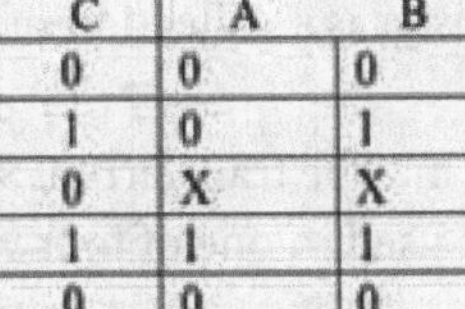

$T_C = AC + A'C'$

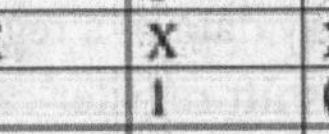

Not self-correcting

This design leads an auto locked condition, so we should avoid it.

To correct the design, assume that the next state of 010 is XX0:

A\BC	00	01	11	10
0	1	0	0	0
1	0	X	1	0

$T_C = AC + A'B'C'$

self-correcting

In this case the counter entered in unwanted states comes back to a valid state. So this design we should consider.

Example 4.3

Show that a BCD ripple counter can be constructed using a 4-bit binary ripple counter with asynchronous clear and a NAND gate that detects the occurrence of count 1010.

Solution:

As soon as the state 1010 is reached, without waiting for the clock it CLEARs the counter to state 0000 with asynchronous CLEAR input given to all FFs simultaneously.

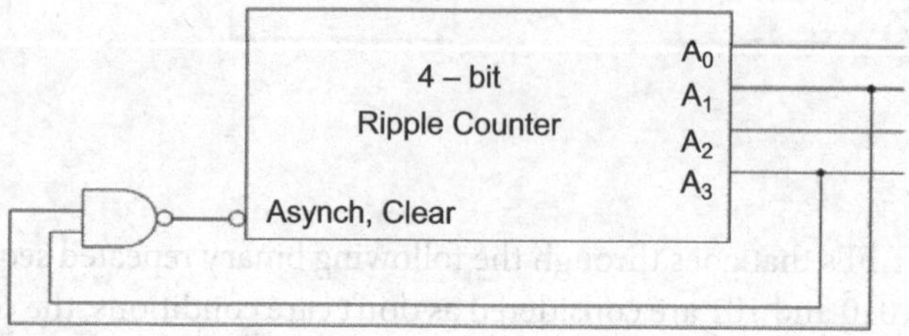

4.5 REGISTERS

FFs are 1 bit memory cell that may be used to store digital information. Using a bunch of FFs the storage capacity in terms of bits can bit increased. A Register is a collection of FFs. An n-bit word can be stored in the n-bit register, which is made up of n FFs.

From one FF to the next, binary data in a register may be transferred. Shift registers are the registers that permit such data transfers. A shift register can be used in one of four ways:

- Serial Input Serial Output
- Serial Input Parallel Output
- Parallel Input Serial Output
- Parallel Input Parallel Output

Shift Registers

4.5.1 Serial Input Serial Output (SISO)

In SISO shift register, data is serially shifted either left side (Left Shift Register) or right side (Right Shift Register). Let's assume that all of the FFs start in the RESET state, which is Qn = ... = Q2 = Q1 = 0. If the n-bit binary number 1....11 is to be entered into the register, the LSB bit should be applied to the Din bit followed by every higher order bit in every next clock cycle. The serial data input Din is linked to the FF's D input. SISO can have two configurations as presented in Fig. 24. If data in (Din) is applied to FF1- at D1 and data output is taken from FFn at Qn it is Right Shift Register. If data in (Din) is applied to FFn - at Dn and data output is taken from FF1 at Q1 it is Left Shift Register. SISO is slowest among all shift registers.

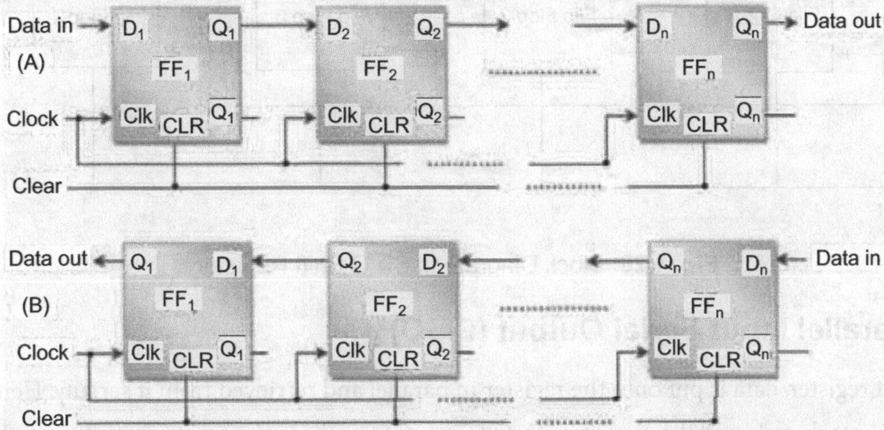

Fig. 4.24: Block Diagram of SISO (A) Right Shift Register, (B) Left Shift Register

SISO Right Shift Register is explained by below waveforms shown in Fig. 4.25. It can be seen that the first bit (LSB) of the data input takes n-clock cycles to get pushed to the output of nth FF. After n-cycles, every input bit can be received in every clock cycle, so MSB will be received after (2n-1) cycles.

4.5.2 Serial Input Parallel Output (SIPO)

In this data is entered serially and extracted simultaneously. Bit by bit, the data is loaded. Outputs are disabled during data loading. After data is successfully loaded all the FF outputs are activated to receive the data at output lines simultaneously or in parallel. An n- bit word must be loaded in n- clock cycles. So the data is available at nth clock cycle only. As a result, the SIPO mode's operating speed is the same as the speed of SISO mode. Fig. 4.26 presents SIPO with serial in and parallel out.

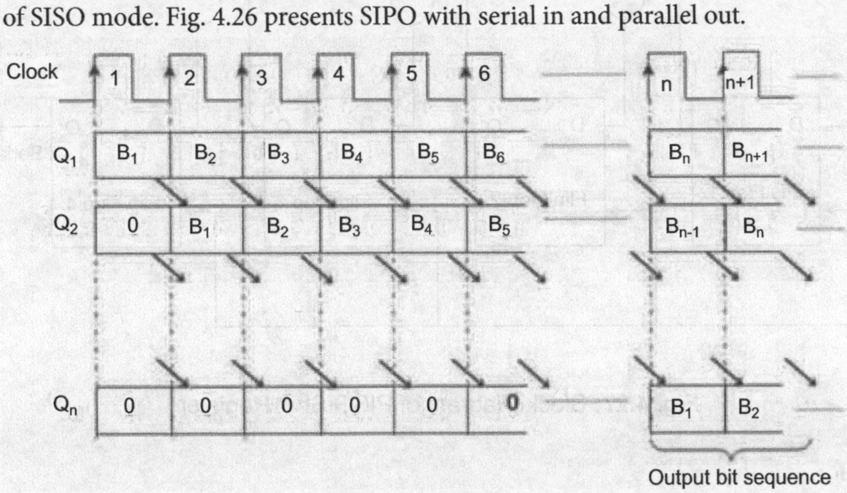

Fig. 4.25: Waveform of SISO Right Shift Register

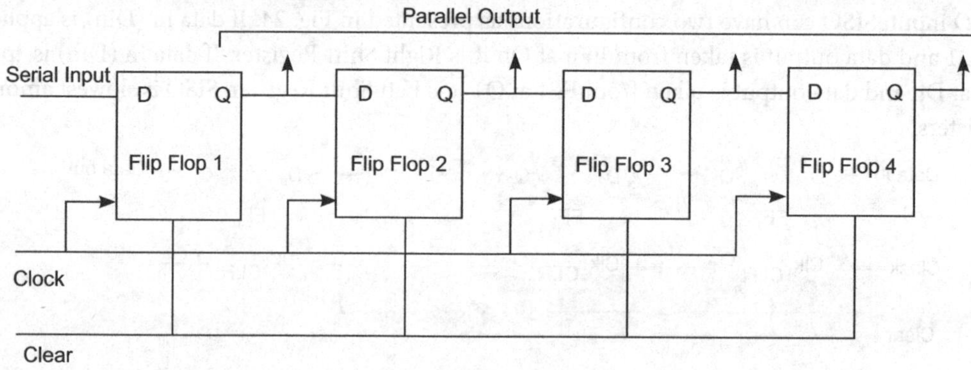

Fig. 4.26: Block Diagram of SIPO Shift Register

4.5.3 Parallel Input Serial Output (PISO)

In PISO shift register, data is put onto the register in parallel and retrieved from it serially. Here the data is entered in 1 clock cycle and the output is started from 1st to nth clock cycle for an n-bit word. Fig. 4.27 presents PISO with 4-bit parallel in serial out. Shift = "0" is used for parallel data load while shift = "1" is used for Serial Right Shifting- SISO.

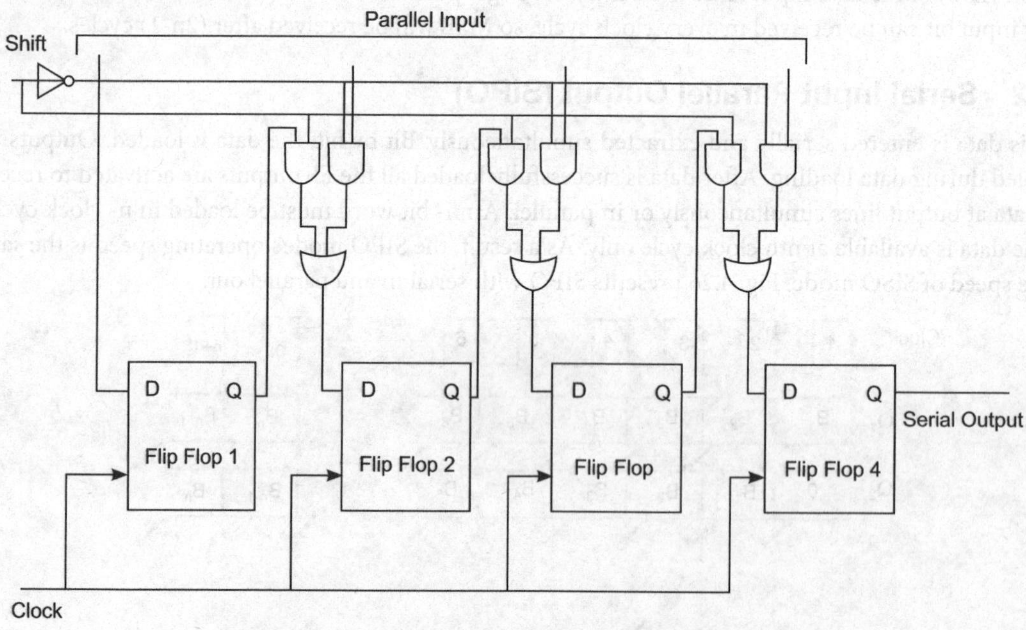

Fig. 4.27: Block Diagram of PISO Shift Register

4.5.4 Parallel Input Parallel Output (PIPO)

In this mode data is entered parallelly and can be seen at the output parallelly, So shifting takes just one clock cycle. Shifting in PIPO is independent on word size. This is fastest shifting among all shift registers. A four bit PIPO is shown in Fig. 4.28. The four FFs data inputs "D" are given binary input bits. The binary input bits will all be fed into the FFs in parallel when a positive clock edge is applied. The output side will see all of the loaded bits at once at outputs of all D-FF as Q1Q2Q3Q4.

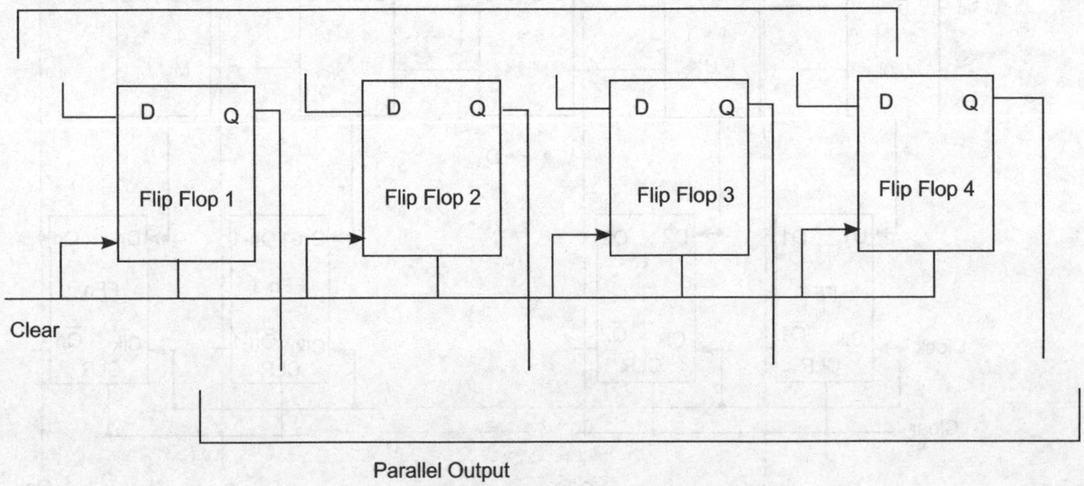

Fig. 4.28: Block Diagram for PIPO Shift Register

4.5.5 Universal Shift Register

This type of register has the ability to be used as SISO, SIPO, PISO or PIPO. Fig. 4.29 presents a n-bit universal shift register. Since there are total 4 modes and at a time any one of these modes is expected to work, so a 4X1 MUX with 2 selection lines is taken.

The mode is selected with S0S1. The different modes can be selected for loading the data serially or parallelly or for left shift or right shift- using S0S1 = 00 to retain the same data, S0S1 = 01 is for serial input for right shift register, S0S1 = 10 - is for serial input for left shit register, S0S1 = 11 - parallel input. Outputs can be obtained parallelly from Q1 to Qn or serially from Q1 (in case of left shit) or Qn (in case of right shit).

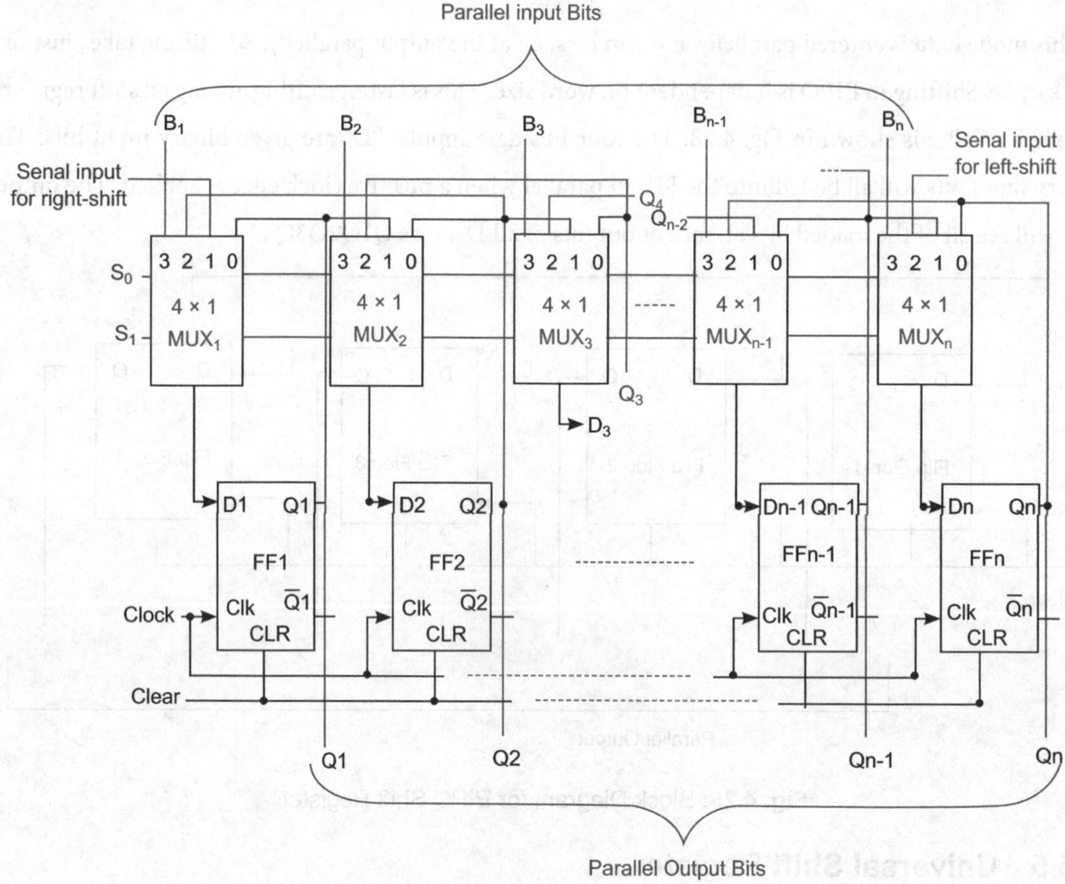

Fig. 4.29: Universal Shift Register

Example 4.4

Design a serial 2's complementer with shift register and a FF. The binary number is shifted out from one side and it's 2's complement shifted into the other side of the shift register.

Solution:

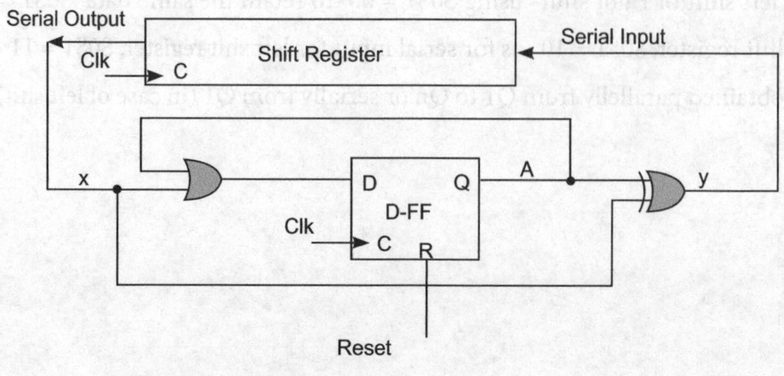

Example 4.5

Disign a 4-bit shift register with parallel load using D FFs. These are two control inputs: shift and load. When shift = 1, the content of the register is shifted by one position. New data is transferred into the register when load = 1 and shift = 0. If both control inputs are equal to 0, the content of the register does not change.

Solution:

General kth FF for register is presented for this. If k=1 Serial input is taken else Ak-1 is taken in below Fig. Bk is the input for parallel load.

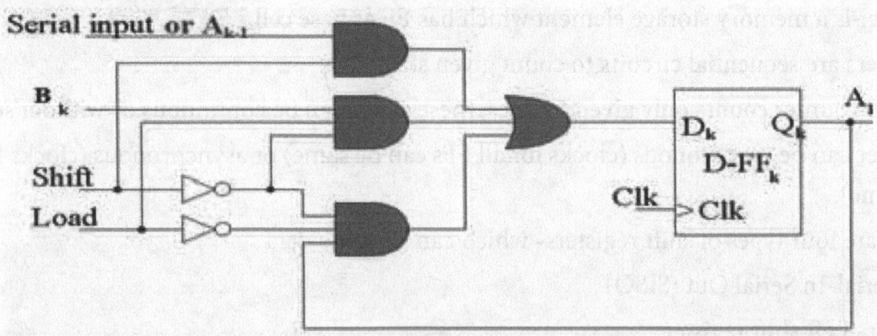

Example 4.6

Draw the logic diagram of a 4-bit register with four D FFs and 4X1 multiplexers with mode selection input s1 and s0. The register operates according to the following function table:

S_1	S_0	Register Operation
0	0	No Change
0	1	Complement the four Output
1	0	Clear register to 0 (Synch)
1	1	Load parallel data

Solution:

Here ith FF of register is shown, Ii is the ith bit of parallel data to be loaded.

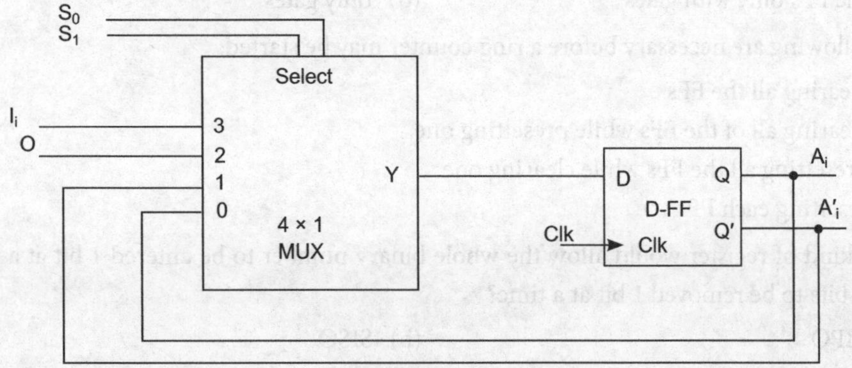

UNIT SUMMARY

- FFs are basic memory cell
- If a latch works with active edge it is know as FF.
- There are 4 types of FFs- SR,JK,T,D
- S-R FF can be converted into D FF by applying S=D and R=D'
- T- flip flp is prepared by tying J-K inputs together.
- Register is a memory storage element which has FF as base cell.
- Counters are sequential circuits to count given states
- Mod -5 counter counts only given 5 states, these states can be continuous or without sequence.
- Counter can be synchronous (clocks for all FFs can be same) or asynchronous (clocks to all FFs are not same).
- There are four types of shift registers- which can shift the data
  - Serial-In Serial Out (SISO)
    - Left shift register
    - Right shift register
  - Serial-In Parallel Out (SIPO)
  - Parallel-In Serial Out (PISO)
  - Parallel-In Parallel Out (PIPO)

EXERCISES

Multiple Choice Questions

1. The speed of a ripple counter is constrained by the delay time of:
   (a) every FF (b) gates and all FFs
   (c) the FFs only with gates (d) only gates
2. The following are necessary before a ring counter may be started:
   (a) clearing all the FFs
   (b) clearing all of the FFs while presetting one.
   (c) presetting all the FFs while clearing one.
   (d) resetting each FF
3. What kind of register would allow the whole binary number to be entered 1 bit at a time and the stored bits to be removed 1 bit at a time?
   (a) PIPO (b) SISO
   (c) SIPO (d) PISO

4. As opposed to ripple counters, synchronous counters have no latency issues because
(a) The first and final stages are the only ones that receive input clock pulses.
(b) Only the last step is given input clock pulses.
(c) None of the counter stages are activated by input clock pulses.
(d) Each stage receives input clock pulses simultaneously.

5. The following is one of the main issues with using asynchronous counters:
(a) Internal propagation delays restrict low-frequency applications.
(b) Internal propagation delays restrict high-frequency applications.
(c) Asynchronous counters are appropriate for use in high- and low-frequency counting applications and don't have any significant limitations.
(d) Asynchronous counters cannot be used in high-frequency applications because they lack propagation delays.

6. The Ring and Johnson counter comparison reveals that:
(a) Ring counter need additional decoding circuitry while having fewer FFs.
(b) Ring counters have an inverted feedback route and
(c) Johnson counter uses fewer decoding circuits but more FFs.
(d) Johnson counter has an inverted feedback route

7. Which form of counter circuit can readily produce a series of evenly spaced timing pulses?
(a) shift register sequencer (b) clock
(c) Johnson (d) binary

8. What does it mean to load the register in parallel?
(a) Concurrently shifting data in all FFs.
(b) Data may be loaded into two of the FFs
(c) Simultaneously loading data into all four FFs
(d) briefly turning off the synchronized RESET and SET inputs

9. What is the name of a shift register that can push data left or right and accepts a parallel input?
(a) tri-state (b) end around
(c) bidirectional universal (d) conversion

10. In an asynchronous binary down counter, when a clock pulse occurs, what occurs to the simultaneous output word?
(a) The last word shrinks by 1 (b) The word output shrinks by 2
(c) The word output grows by 1 character (d) The last word grows by

11. The most typical applications for mod-6 and mod-12 counters are:
(a) frequency counters (b) multiplexed displays
(c) digital clocks (d) power consumption meters

12. The SIPO shift register sends data over a single line, one bit at a time, from one parallel data bus to another parallel data bus
(a) True (b) False

13. One bit at a time is transferred from one line of a concurrent bus to another line using a SISO register.
(a) YES (b) NO

14. A ripple counter is an asynchronous counter.

(a) True (b) False

15. Counting devices are often used in digital clocks.

(a) YES (b) NO

16. Each state of an asynchronous counter is timed by the same pulse.

(a) True (b) False

17. All data bits are simultaneously entered into a parallel-in, serial-out shift register, which sends each bit out one at a time.

(a) True (b) False

18. A counter's modulus (mod) is equal to its highest count (N).

(a) True (b) False

19. Consider the partial implementation of a 2-bit counter using T FFs following the sequence 0-2-3-1-0, as shown below

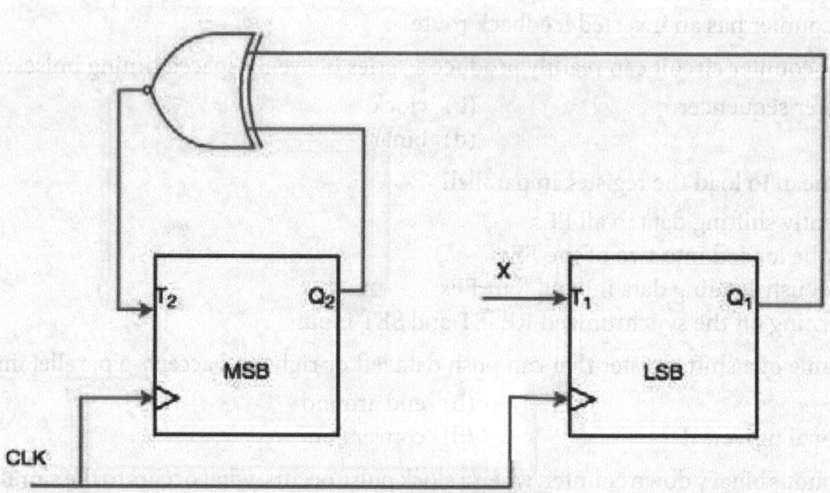

To complete the circuit, the input X should be

(a) Q2? (b) $Q_2 + Q_1$

(c) $(Q_1 \text{ XOR } Q_2)'$ (d) $(Q_1 \text{ XOR } Q_2)$

20. A four-bit binary counter has control signals which are functioning as per the below table. The implementation of counter is as per below diagram (see on next page)-

Assume that the counter and gate delays are negligible. If the counter starts at 0, then it cycles through the following sequence:

(a) 0, 3, 4 (b) 0, 3, 4, 5

(c) 0, 1, 2, 3, 4 (d) 0, 1, 2, 3, 4, 5

Clear	Clock	Load	Count	Function
1	×	×	×	Clear to 0
0	×	0	0	No change
0	↑	1	×	Load Input
0	↑	0	1	Count next

The counter is connected as follows:

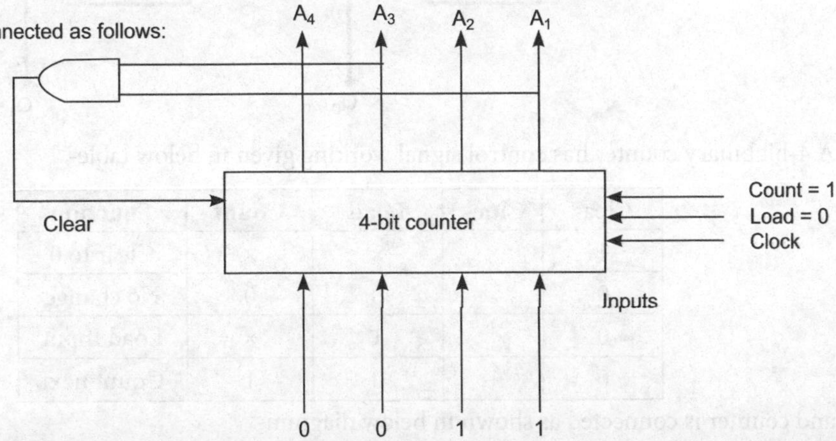

ANSWERS															
1.	(a)	2.	(b)	3.	(b)	4.	(d)	5.	(b)	6.	(d)	7.	(a)	8.	(c)
9.	(c)	10.	(a)	11.	(c)	12.	(b)	13.	(b)	14.	(a)	15.	(a)	16.	(b)
17.	(a)	18.	(a)	19.	(d)	20.	(c)								

Short and Long Answer Type Questions

1. What is a ripple counter?
2. What is a mod-5 counter?
3. How is counter value calculated? (Hint: counter states, FF)
4. How to design the mod 3 counter using 3 FFs? (Hint: one hot encoding)
5. What can be the modulus of a 3 FF counter, explain with reason?
6. Write the truth tables and state diagrams for both a J-K FF and a D FF.
7. Two NAND gates are cross couped to design an SR latch, if both S-R inputs are 0, find the outputs.
8. Design active HIGH inputs SR latch using NOR gates (only)and NAND gates (only). Compare both the designs.

Numerical Problems

1. In a three stage counter, using RS FFs what will be the value of the counter after giving 9 pulses to its input? Assume that the value of counter before giving any pulses is 1.
2. We want to design a synchronous counter that counts the sequence 0-1-0-2-0-3 and then repeats. What is the minimum number of J-K FFs required to implement this counter?

3. Design a synchronous counter to go through the following states: 1, 4, 2, 3, 1, 4, 2, 3, 1, 4,..........
4. If Q_1Q_0 is 00 initially, find the next four outputs of Q_1Q_0?

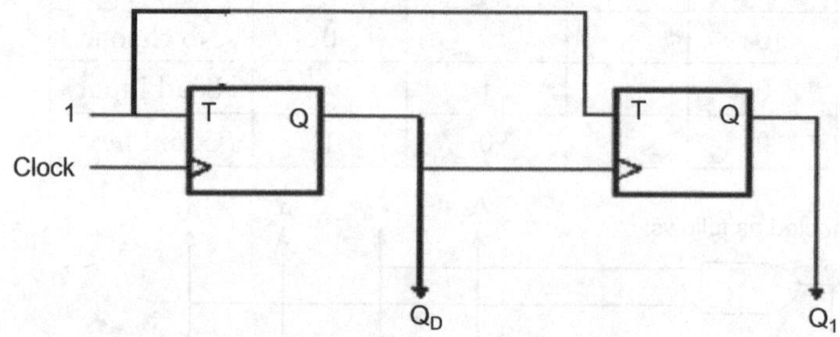

5. A 4-bit binary counter has control signal working given in below table-

Clear	Clock	Load	Count	Function
1	×	×	×	Clear to 0
0	×	0	0	No change
0	↑	1	×	Load Input
0	↑	0	1	Count next

and counter is connected as shown in below diagram-

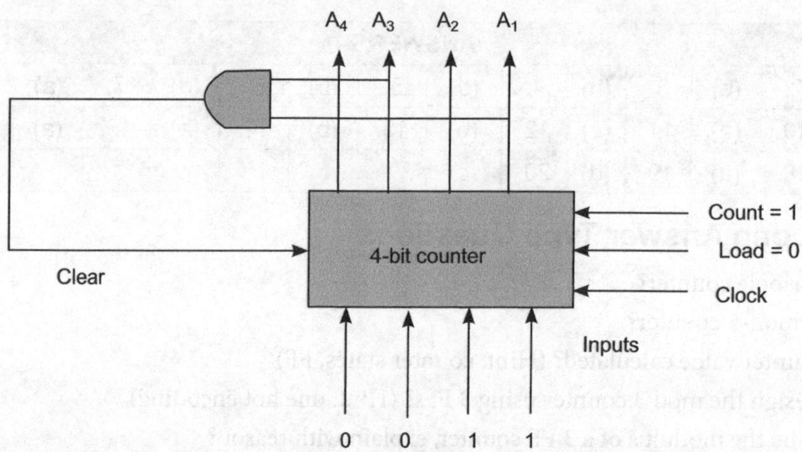

Find the states for counter if it starts at 0 and all delays are negligible.

6. Complete the two-bit counter using T-FFs which should have 0->2->3->1->0 sequence, find X.

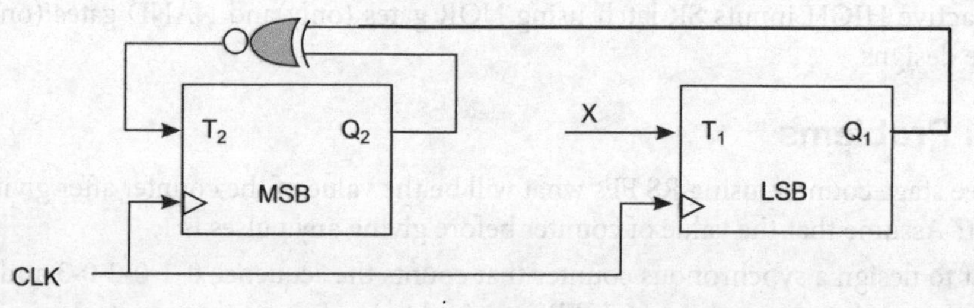

7. Assume initial state $Q_0Q_1 = 00$. The D FFs have set up times 20 ns (nanosecond) and hold time 2 ns. For the timing diagrams given here for X and C; the clock period of C <= 40 ns (nanosecond), find plot of Y output?

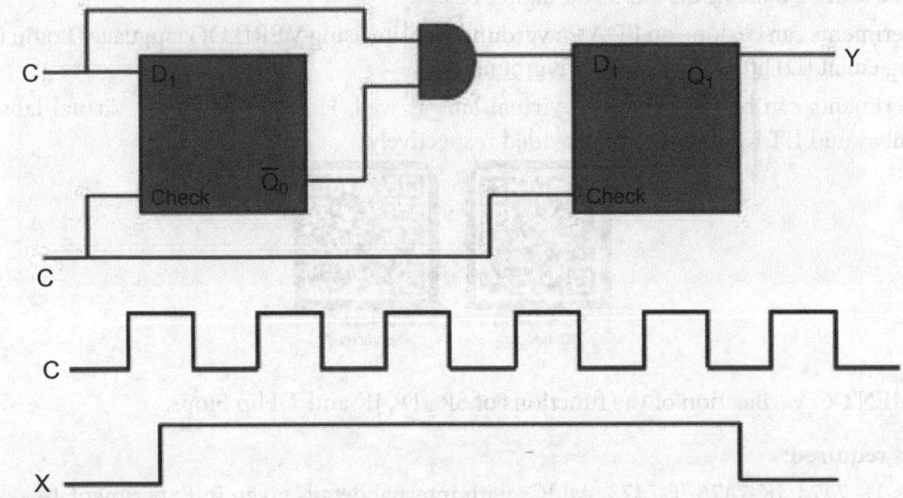

8. The below sequential circuit has initial state $Q_0Q_1Q_2$= 100. General state for the circuit is presented by $4Q_2 + 2Q_1 + Q_0$. Find the state sequence of the circuit?
(Ans:- mod -8 counter with sequence-> 4-1-2-5-6-7-3-1-4)

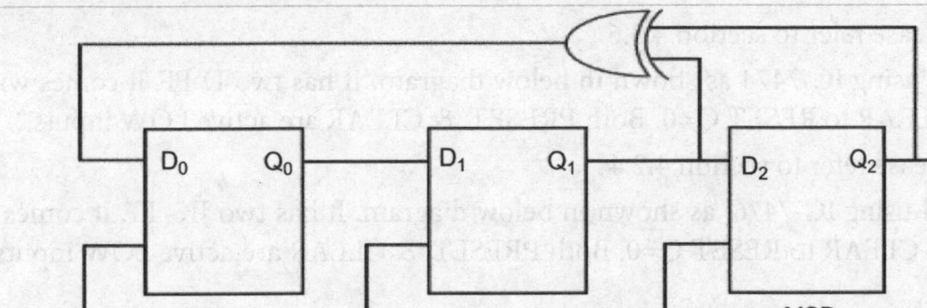

9. Consider initial state of a four bit Johnson counter 0000. Find the counting sequence for this counter.

10. Initial state of the below master-slave FF is PQ= 01. Find the output after four clock cycles.

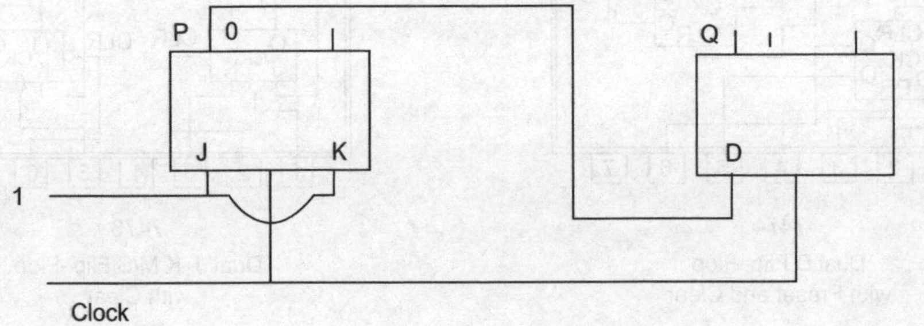

PRACTICAL

Experiments can be done in the lab using digital ICs.

— Experiments can be done on EDA Playground website using VERILOG language (Login is required using email ID).https://www.edaplayground.com

— Experiments can be performed in virtual labs as well. Here QR code for Virtual labs from IIT Bombay and IIT Kharagpur are provided respectively.

Virtual-Lab IIT Kharagpur

Virtual-Lab IIT Bombay

EXPERIMENT 6. Verification of the functions of SR , D , JK and T Flip Flops.

Apparatus required:

Digital ICs-D- 7474, JK-7476, T-7473. All ICs with internal details given in Experiment 1.

Diagram and Truth Table

SR Flip Flop-Please refer to section 4.2.1

SR FF can be designed using back to back connected NAND gates as shown in Fig. 4.2 and verify Table 4.1(a) for active LOW inputs.

D Flip Flop – Please refer to section 4.3.6

D-FF is verified using IC-7474 as shown in below diagram. It has two D FF. It comes with PRESET to SET Q=1 and CLEAR to RESET Q=0. Both PRESET & CLEAR are active LOW inputs.

JK Flip Flop- Please refer to section 4.2.4

JK-FF is verified using IC-7476 as shown in below diagram. It has two JK- FF. It comes with PRESET to SET Q=1 and CLEAR to RESET Q=0. Both PRESET & CLEAR are active LOW inputs. Anoth Ic for JK-FF is 7473.

T Flip Flop- Please refer to section 4.2.7

T-FF can be prepared using JK-FF IC and by tying both inputs J &K together. Either one IC of JK-FF 7476 or 7473 can be used. for IC 7473 comes with active LOW CLEAR input to RESET output Q=0.

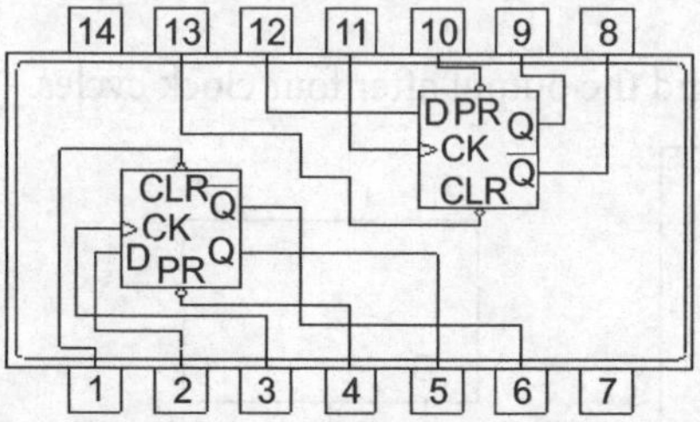

7474

Dual D Flip–Flop with Preset and Clear

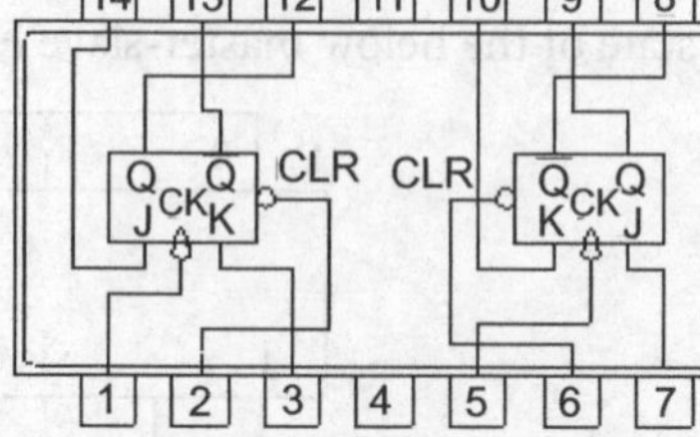

7473

Dual J–K M/S Flip–Flop with Clear

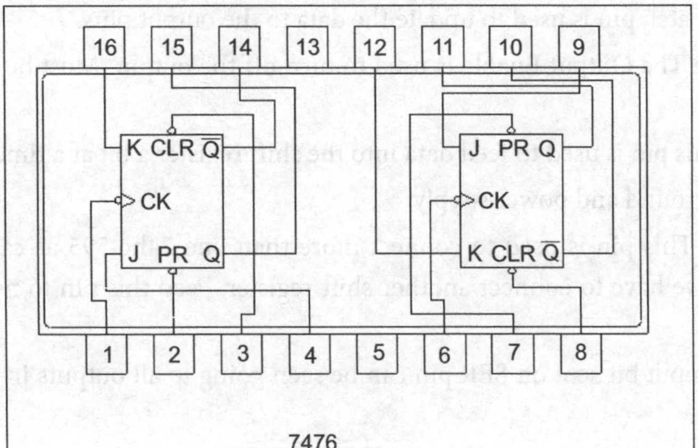

7476
Dual J–K M/S Flip–Flop
with Present and Clear

Observations: Complete the truth table of each flip flop by applying proper input pattern and observing the outputs and verify its correctness.

Conclusion: S-R, JK, D, T flip flops' functions have been verified using digital ICs.
Visit wepage for- Verilog code for D Flip Flop - FPGA4student.com

EXPERIMENT 7: Design Controlled Shift registers.

Apparatus: Shift Register IC 74HC595

IC-74HC595 working

Details of Shift Register Pins:

Q_A to Q_H are the 8 output pins and shoudl be connected to some type of output like LEDs, 7 segments etc.

(SRCLR') Master Reset: Resets all outputs as LOW. Must be held high for normal operation.

(SRCLK) Clock: This is the clock pin to which the clock signal has to be provided from MCU/MPU

(RCLK) Latch: The Latch pin is used to update the data to the output pins.

(OE') Output Enable The Output Enable is used to turn off the output. Must be held LOW for normal operations.

(SER) Serial Data This pin is used to feed data into the shift register a bit at a time.

GND and VCC are ground and power supply

(Q_H) Serial Output: This pin is used to connect more than one 74hc 595 as cascading (Pin 9 is data out and used when we have to connect another shift register, Feed this pin to next Shift register IC, if connected)

Observations: The input bit sent on SER pin can be seen going to all outputs in every next clock cycle sequentially.

Conclusion: Left shift register operations are verified using 74HC595 IC.

More about IC 74HC595

EXPERIMENT 8. To design a programmable Up-Down Counter with a 7-segment display

Apparatus: IC7476, 7486, connecting wires

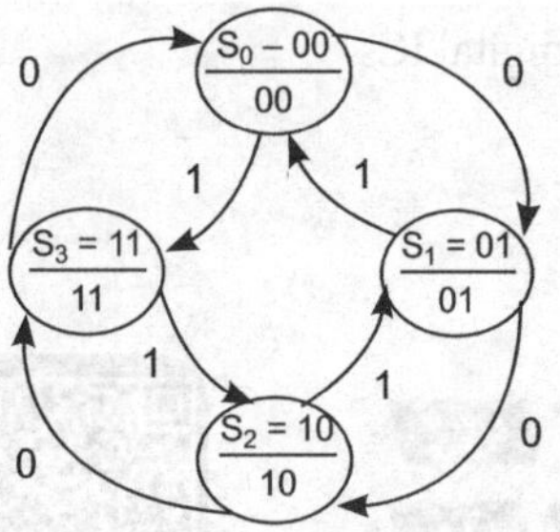

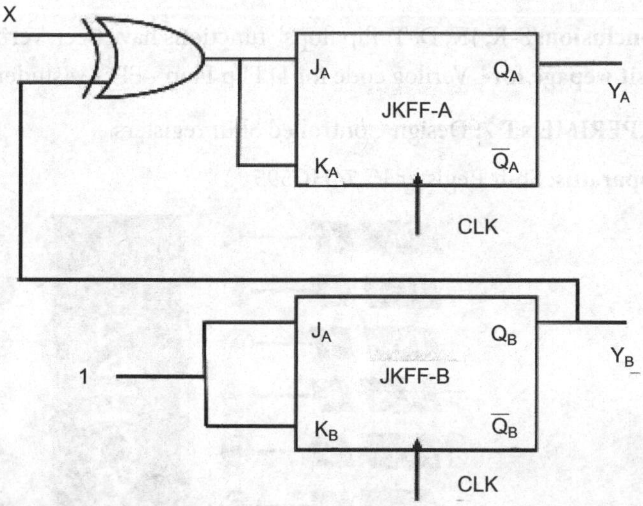

$Y_A = Q_A$ and $Y_B = Q_B$

$J_A = K_A$ and $Q_B \oplus X$

$J_B = K_B = 1$

As represented by the state diagram here input X-acts as UP COUNTER mode when X=0 and DOWN COUNTER mode when X=1.

With the help of JK- FF IC 7476 or 7473 and XOR gate IC 7486 it can be designed as the connections done in the above logic diagram.

Observations: Complete the state diagram for X=0 and X=1 input patterns and observing the outputs and verify its correctness for UP Counter and DOWN Counter respectively.

Conclusion: UP and DOWN counters' functionalities have been verified using digital ICs.

Verilog code for Counter

Verilog designs and testbench for D flipflop-

D Flip Flop Design

```
module D_FF(D,clk,sync_reset,Q);

input D; // Data input
input clk; // clock input
input sync_reset; // synchronous reset
output reg Q; // output Q

always @(posedge clk)
begin
 if(sync_reset==1'b1)
  Q <= 1'b0;
 else
  Q <= D;
end

endmodule
```

D -FF Test Bench

```
`timescale 1ns/1ps;
module tb_DFF();
reg D, clk, reset;
wire Q;
D_FF dut (D,clk,reset,Q);
initial begin
 clk=0;
   forever #10 clk = ~clk;
end
initial begin
 reset=1;
 D <= 0;
 #100;
 reset=0;
 D <= 1;
 #100;
 D <= 0;
 #100;
 D <= 1;
end
endmodule
```

KNOW MORE

Before the semiconductor based memories specially flip flop based, the data was stored in magnetic memory. Magnetic memory examples are tapes which were available to record both audio and video. Optical memory devices like Compact Disk (CD) and DVDs were also available to store the data. Now a days pen drive, hard disk SD card are available to store very big size of data.

REFERENCES AND SUGGESTED READINGS

1. M. Morris Mano and Michael Ciletti, "Digital Design", 5th edition, Pearson.
2. Digital Electronic Circuits - Course (nptel.ac.in)

Dynamic QR Code for Further Reading

-QR codes are embedded in the chapter with the topics.

□□

5 Memory Devices

UNIT SPECIFICS

Through this unit we have discussed the following aspects:

- General Classification of memories;
- RAM and ROM organization;
- Types of different RAMs and ROMs;
- Advantages and Disadvantages of different memories;

The practical applications of the topics are discussed for generating further curiosity and creativity as well as improving problem solving capacity.

Besides giving a large number of multiple choice questions as well as questions of short and long answer types marked in two categories following lower and higher order of Bloom's taxonomy, assignments through a number of numerical problems, a list of references and suggested readings are given in the unit so that one can go through them for practice. It is important to note that for getting more information on various topics of interest some QR codes have been provided in different sections which can be scanned for relevant supportive knowledge.

After the related practical, based on the content, there is a "Know More" section. This section has been carefully designed so that the supplementary information provided in this part becomes beneficial for the users of the book. This section mainly highlights the initial activity, examples of some interesting facts, analogy, history of the development of the subject focusing the salient observations and finding, timelines starting from the development of the concerned topics up to the recent time, applications of the subject matter for our day-to-day real life or/and industrial applications on variety of aspects, case study related to environmental, sustainability, social and ethical issues whichever applicable, and finally inquisitiveness and curiosity topics of the unit.

RATIONALE

This fundamental unit focuses on computer hardware fundamentals with the overview of various types of memories used in computers. The unit begins with the classification of memories and then followed by the description of each type of memory. It explains about the operations of various types of memories and their applications. It also explains how to calculate the size of a memory using given data.

As the scope of digital electronics extend to domains such as healthcare, software, automation, digital banking, smart watches, smart phones and many more, various memories are used to store the data in computers. These are important part of any digital electronics system or computer hardware and can be found in CDs, digital camera, DVDs, pen drives, MP3 players and many more portable electronics.

PRE-REQUISITES

Basic concept of logic gates,

UNIT OUTCOMES

List of outcomes of this unit is as follows:

U5-O1: Understand type of memories

U5-O2: Learn operation of memories

U5-O3: Learn to calculate the size of a memory

U5-O4: Understand the structure of different memories using logic gates

U5-05: Understand the use of memory at appropriate place in CPU

Unit-5 Outcomes	EXPECTED MAPPING WITH COURSE OUTCOMES (1 - Weak Correlation; 2 - Medium Correlation; 3 - Strong Correlation)					
	CO-1	CO-2	CO-3	CO-4	CO-5	CO-6
U5-O1	-	2	3	3	3	-
U5-O2	-	2	3	3	3	-
U5-O3	-	2	3	3	3	-
U5-O4	-	2	3	3	3	-
U5-O5	-	2	3	3	3	-

"Education is the most powerful weapon you can use to change the world."

–BB King

5.1 INTRODUCTION

A "memory" is an important part of "computer" and any other "electronic system". Memory is used to store data, program and instructions used for executing any program. It can be used for temporary storage or permanent storage. Memory is a part of Central Processing Unit (CPU) of computer. A "CPU" is having "Arithmetic Logical Unit (ALU)", "Control Unit (CU)" and "Memory Unit". The general block diagram of computer system with memory is shown in Fig. 5.1.

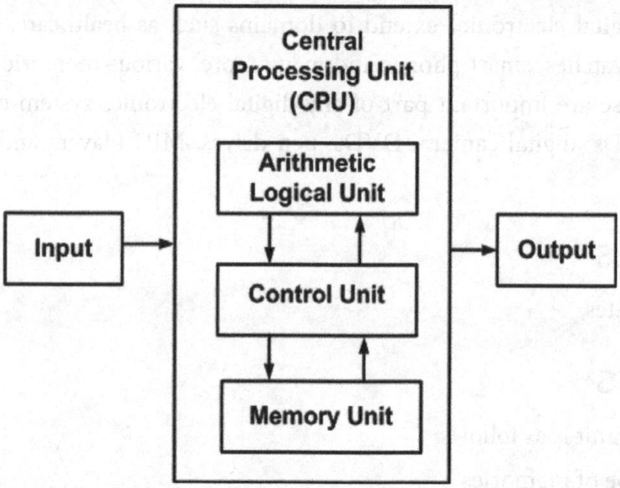

Fig. 5.1: Block diagram of Computer system with memory unit.

5.2 CLASSIFICATION OF MEMORIES

The classification of Memory is given in Fig. 5.2. There are two broad types of memory-

1. Primary Memory- It is Internal memory.
   - "RAM"
     - "Static RAM (SRAM)", "Dynamic RAM (DRAM)"
   - "ROM"
     - PROM, EPROM, EEPROM
2. Secondary Memory-It is External memory.
   - Hard disk
   - Pendrive

Other memory :

3. Cache Memory- It is Internal memory within processor.
4 Register Memory

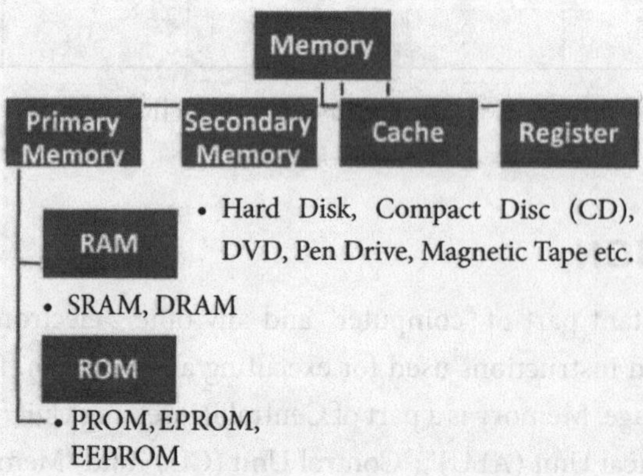

Fig. 5.2: Classification of Memory.

5.3 MEMORY STRUCTURE: ADDRESS AND SIZE

"Memory" is a collection of "storage cells", which stores data and information in group of bits know as words. A word is a sequence of 0s and 1s and each word is stored in a memory location, known as address. The group of eight bits is knows as a byte and the total number of bytes stored by memory is known as capacity of memory or memory size.

The general "block diagram" of memory unit is shown in Fig. 5.3. It consists of "m address lines" that means it can access "2m locations/addresses". It also has "n data lines" (input/output lines) through which data can be transferred in and out of the memory. The "Read" and "Write" ports of the "memory" are called "control lines" or "control inputs". The Write port causes binary data on data lines to be transferred into the memory and the Read port causes the binary data to be transferred out of the memory from a memory location.

The general organization of a memory device is shown in Fig. 5.4. An address decoder selects any one row of memory array to be written or read from. For example, if (101)2=(5)10 is placed on the address lines, the fifth row of the memory will be selected and connected to the data lines (considering first row is the Row 1).

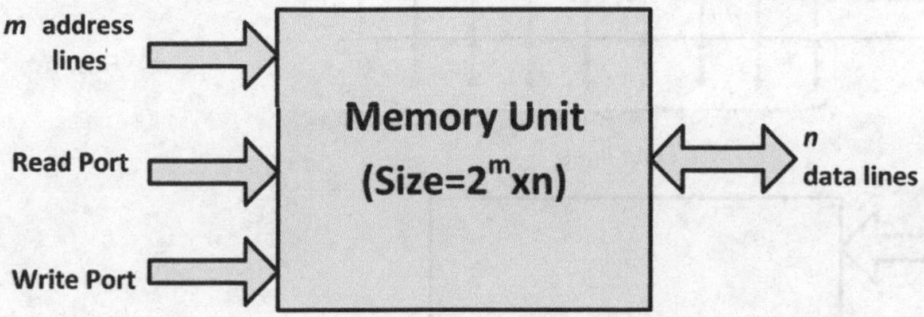

Fig. 5.3: Basic structure of a memory unit.

The Write enable and Read Enable specify whether the data is being written to or reading from the selected row of memory array. These signals can also be written as "$R/\overline{W}$"; here "$R/\overline{W}$=1" indicates data read and "$R/\overline{W}$=0" indicates data write.

The Chip select is an active low signal which is used to enable or disable the memory device. It can be represented as $\overline{CS}$; if $\overline{CS}$=0, the memory becomes idle (disconnected from data input and output lines) and if $\overline{CS}$=1, the memory enables all the input and output lines for data transfer.

5.4 RANDOM ACCESS MEMORY (RAM)

"Random Access Memory (RAM)" is a hardware device in a "computer" system that "stores the data" and "programs" to be accessed quickly by the processor. It is volatile memory that means it cannot store data permanently; it is lost when the computer is turned OFF and it is only retained in the RAM as long as the "computer" is ON. "RAM" is primary memory of computer and it is also known as Read-Write Memory (RWM) because it can perform both Read and Write operations.

The internal organization of RAM is shown in fig. 5.6. It consists of register banks (B0, B1, B2, and B3), bit-addressable area and scratch-pad area.

Fig. 5.4 can also be represented as general memory organisation block diagram as shown in Fig. 5.5.

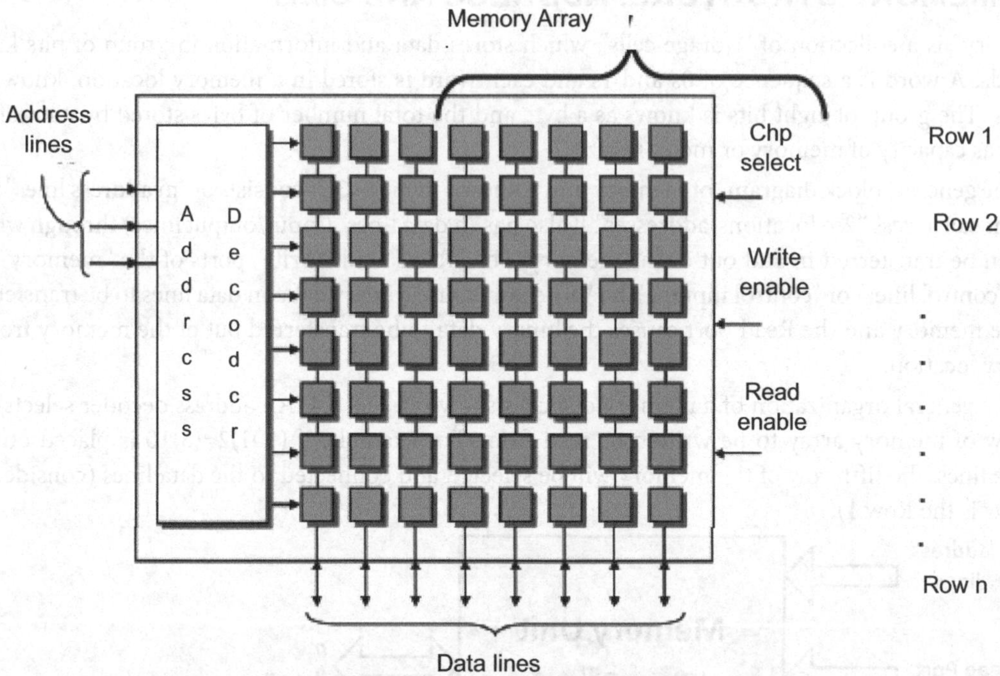

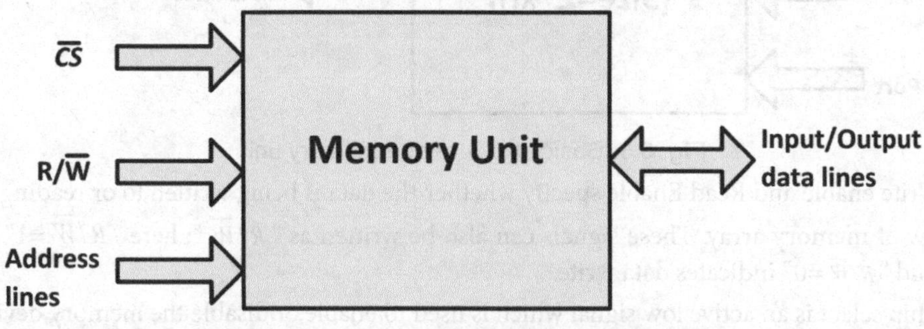

Fig. 5.4: General Organization of Memory Device.

Register Banks: The RAM has four register banks (B0, B1, B2 and B3) and each contains 8-bit general-purpose registers R0-R7. The memory locations of these registers are shown in Fig. 5.6.

Bit-addressable area:

It consists of "bit-addressable registers" and used to store the "bit variables" from "application program" such as status of any "output device." It has 128 addresses from 00h to 07Fh that represents the location of data storage. These are designed from the addresses 20h to 2Fh.

Scratch-pad area:

It consists of byte-addressable registers and used to store the byte variables from application program. These are designed from the addresses 30h to 7Fh.

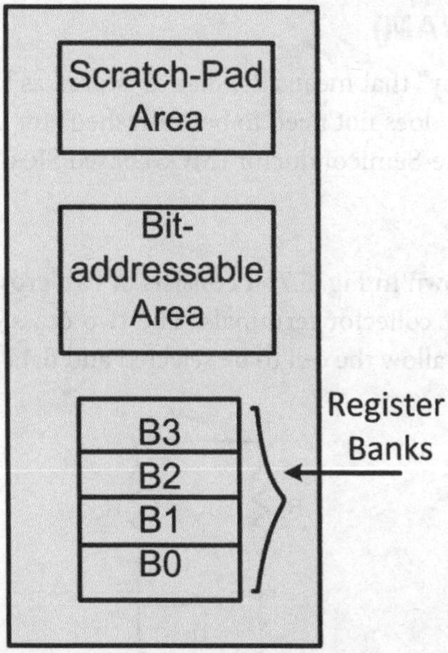

Fig. 5.5: Internal organization of RAM.

Know more
RAM

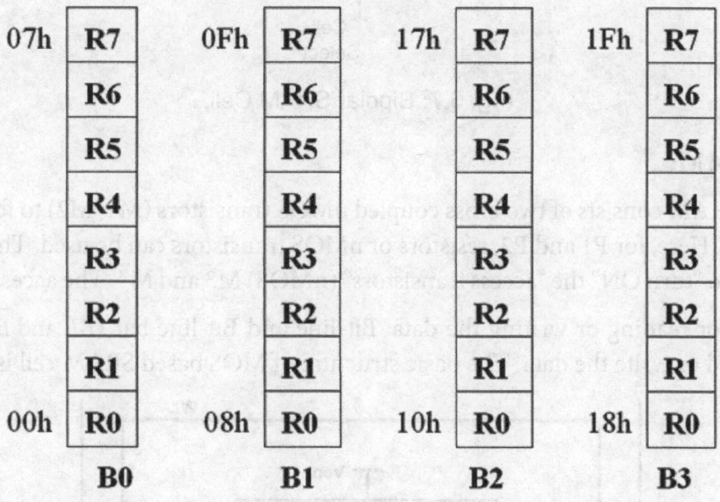

Fig. 5.6: Register banks and their memory locations in RAM.

5.4.1 Types of RAM

RAM is classified into Static RAM, Dynamic RAM. Static RAM can be bipolar or MOS flip-flop and Dynamic RAM is a type of MOS family. Semiconductor memories have large storage capacity and used in most of the electronic/computer systems.

5.4.2 Static RAM (SRAM)

An SRAM is a "volatile memory" that means the data is "stored as long as the power supply" is applied. The term "static" means that it does not need to be "refreshed" for updating the data. It is of two types: bipolar SRAM and Metal-Oxide-Semiconductor (MOS) based SRAM. SRAM is used in cache memory.

Bipolar SRAM:

The basic bipolar SRAM is shown in Fig. 5.7. It consists of two cross coupled bipolar transistors (BJTs) and two resistors connected at collector terminals. The two cross coupled transistors form a bistable latch. The emitter connections allow the cell to be selected and data to be read out.

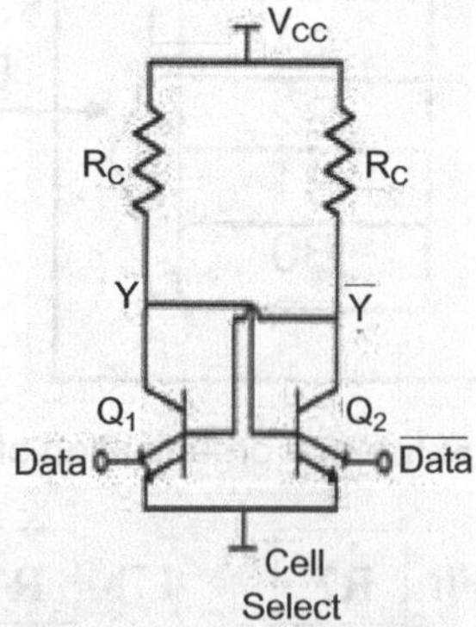

Fig. 5.7: Bipolar SRAM Cell.

MOS based SRAM:

The MOS based SRAM consists of two cross coupled nMOS transistors (M1, M2) to form a bistable latch and P1, P2 as load. Here, for P1 and P2, resistors or nMOS transistors can be used. The WL is the "Word Line" that is used to "turn ON" the "access transistors" (nMOS) M3 and M4. The access transistors enable the memory cell for reading or writing the data. Bit-line and Bit-line bar (*BL* and $\overline{BL}$) act as input or output lines to read or write the data. The basic structure of MOS based SRAM cell is shown in Fig. 5.8.

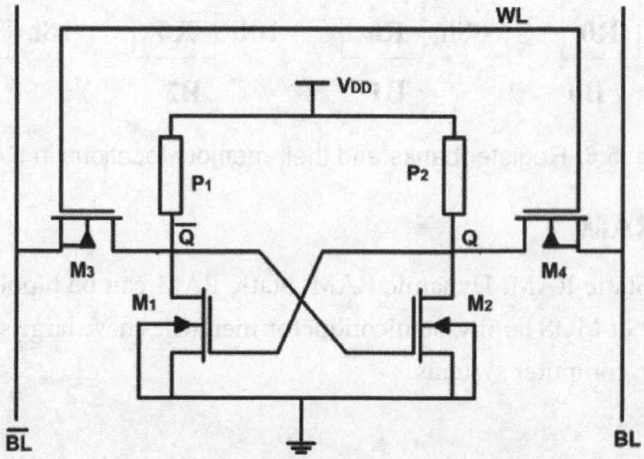

Fig. 5.8: MOS based SRAM Cell.

CMOS based SRAM:

The Complementary MOS (CMOS) based "SRAM cell" is shown in Fig. 5.9. It consists of a "cross-coupled inverters" (combination of PMOS (P1, P2) and NMOS (N1, N2)) and two access transistors (N3, N4).

As shown in Fig.5.9, it has "two cross coupled inverters" (P1-N1 and P2-N2) and "two access transistors" (nMOS-N3, N4). The *BL* and $\overline{BL}$ are used as input lines to read the data from the memory cell and as output lines to write the data in the memory cell. There are 6 transistors in this SRAM, so it is called a 6T SRAM cell. Other variants based on the number of transistors are also available like 7T, 8T SRAM cell etc.

SRAM Working-

To perform Read/Write operation in SRAM cell, Word line –WL is kept HIGH.

Read Operation- To read bit stored in SRAM cell **memory should have some value. First, precharge** BL-bit line and bit bar line- $\overline{BL}$ HIGH by applying HIGH voltage at both lines, once precharge is done remove the HIGH voltages from both Bit lines. Now, to read apply HIGH voltage (Vdd) at WL-line.

Case 1.-Assume Q=1 and Q'=0 is stored in the memory. By making $\overline{BL}$ /BL and WL HIGH makes N4/ N3 "ON". Therefore, these transistors work like pass transistors. For N3 transistor both Q and BL are HIGH, so it will remain HIGH only but for N4, Q'=0 but $\overline{BL}$ line is HIGH, therefore, transistor N4, is ON till $\overline{BL}$ is brought down to LOW voltage. BL and $\overline{BL}$ are connected to the sense amplifier, this sense amplifier acts as a comparator, so When *BL* is low the output will be "1". Hence input Q=1 and we got the output as 1, read operation verified.

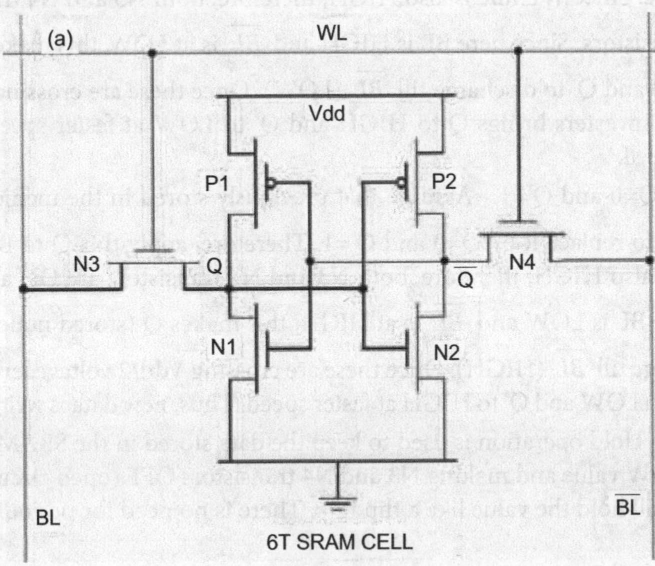

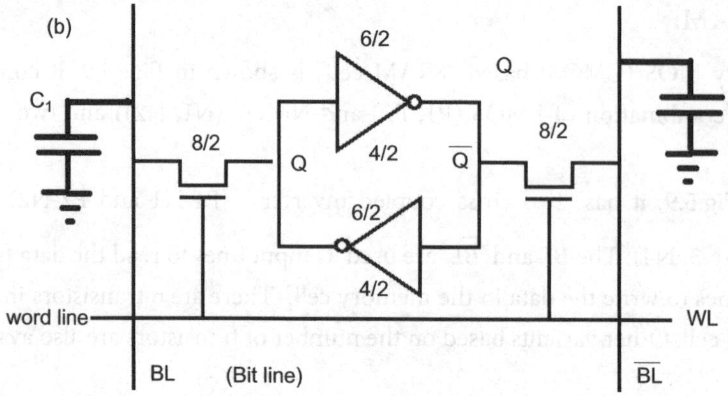

Fig. 5.9: (a) CMOS based SRAM Cell (b) simplified structure with capacitors at bit lines.

Case 2.- Assume Q=0 and Q'=1 is stored in the memory. By making $\overline{BL}$ /BL and WL HIGH makes N4/ N3 "ON". Therefore, these transistors work like pass transistors. For N4 transistor both Q' and $\overline{BL}$ are HIGH, so it will remain HIGH only but for N3, Q=0 but BL line is HIGH, therefore, transistor N3 is ON till BL is brought down to LOW voltage. BL and $\overline{BL}$ are connected to the sense amplifier, this sense amplifier acts as a comparator, so When BL is low the output will be "0". Hence input Q=0 and we got the output as "0", read operation verified.

The PMOS/NMOS ratio must be good enough to bring the node voltages below threshold voltage. This is the read constraint. NMOS has to be stronger to pull down the output to "0" while PMOS has to be stronger to pullup the output to Vdd.

Write Operation- To write in the memory cell make WL line HIGH.

Case 1.- Writing Q=1 and Q'=0 - Assume that previously stored in the memory is Q=0 and Q'=1 and now it is needed to replace it by new data - Q=1 and Q'=0. Therefore, apply this new Q to BL line and new Q' to $\overline{BL}$ line. Since WL line is also HIGH, therefore, both N3 and N4 transistors are ON and working like pass transistors. Since here BL is HIGH and $\overline{BL}$ is at LOW, this makes Q (stored node) to charge till BL (HIGH) and Q' to discharge till $\overline{BL}$ (LOW). Once these are crossing Vdd/2 voltage level, the cross coupling of inverters brings Q to HIGH and Q' to LOW at faster speed. Thus, new data is written in the SRAM cell.

Case 2. Writing Q=0 and Q'=1 - Assume that previously stored in the memory is Q=1 and Q'=0 and now it is needed to replace it by Q=0 and Q'=1. Therefore, apply this Q to BL line and Q' to $\overline{BL}$ line. Since WL line is also HIGH, therefore, both N3 and N4 transistors are ON and working like pass transistors. Since here BL is LOW and $\overline{BL}$ is at HIGH, this makes Q (stored node) to discharge till BL (LOW) and Q' to charge till $\overline{BL}$ (HIGH). Once these are crossing Vdd/2 voltage level, the cross coupling of inverters brings Q to LOW and Q' to HIGH at faster speed. Thus, new data is written in the SRAM cell.

Hold Operation- Hold operation is used to keep the data stored in the SRAM cell. This is done by putting WL line to LOW value and making N3 and N4 transistors OFF (open circuit). The back to back connected inverters will hold the value like a flip flop. There is no need for periodic refreshing the data in SRAM memory cell.

5.4.3 Dynamic RAM (DRAM)

In 1966, Robert Dennard at IBM invented "Dynamic Random Access Memory (DRAM)". DRAM works much differently than other types of memory. The "DRAM" is a type of "semiconductor memory" which consists of a transistor and a capacitor. It "stores" each "bit of data" in a "capacitor". It works on charging/discharging of the capacitor and provides states logic '1'/logic '0' accordingly. It requires periodical refresh operation as the charge stored in the capacitor leaks after some time. The general structure of DRAM is shown in Fig. 5.10, it is a very simple structure as it has only one transistor and one capacitor.

The wordline is connected to gate terminal of the nMOS transistor for controlling the access to the capacitor. If wordline is '1', then the access transistor will turn ON and the capacitor will start charging or discharging based on the voltage at bitline.

The bitline is connected to source terminal of the transistor, it is used to read the charge stored in the cell and also to provide the voltage for writing new value to the cell.

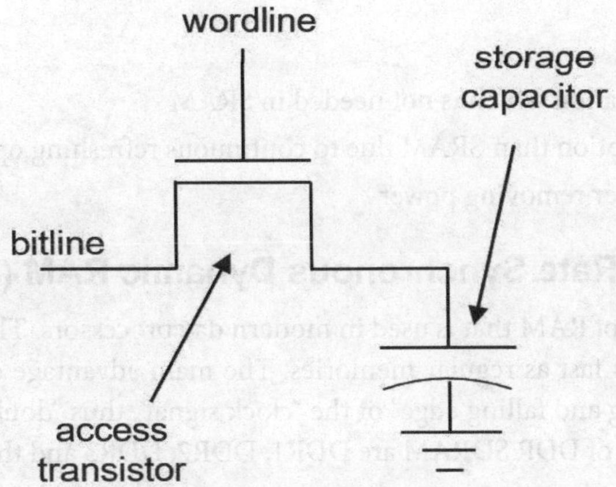

Fig. 5.10: DRAM cell.

DRAM Working-

To perform any operation (Read/Write) in DRAM cell, Word line –WL is kept HIGH.

Read Operation- To read, bit stored in DRAM cell **memory should have some value in the form of charge on capacitor (Charge Q=C.V). First, pre-charge** BL-bit line (using Bit line capacitance) to $V_{dd}/2$ and then put word line-WL to HIGH thus making the access transistor ON (short) and the BL capacitance either get charged more (if storage capacitor was at V_{dd}) or gets discharged through access transistor (if storage capacitor was at "0"). This actually changes the charge on BL capacitor and this change is sensed by sense amplifiers to detect whether "1" bit was stored (capacitance at BL will be charged to higher voltage than $V_{dd}/2$) or "0" bit was stored (capacitance at BL will be discharged to lower voltage than $V_{dd}/2$). Note sense amplifier is a basic comparator that compared the bit line with Vdd/2.

Write Operation- To write in the memory cell make WL line HIGH. Put bit to be written on BL line. WL line HIGH makes the access transistor ON and the storage capacitor gets charged to V_{dd} or discharged to "0" according to BL line is kept HIGH (to write "1") or LOW (to write "0").

Hold Operation- Hold operation is used to keep the data stored in the DRAM memory cell. This is done by putting WL line to LOW value and making access transistor OFF (open circuit). This keeps the storage capacitor to its previous storage value since no charging or discharging path through access transistor. Since there is a subthreshold leakage through access transistor (MOSFET), thus after some time the capacitor gets discharged, so it needs refreshing the DRAM memory data periodically.

Advantages:

- Simple structure
- Low cost than SRAM
- Provides high density than SRAM
- More data can be stored due to higher density
- DRAM is used in main memory while SRAM is used in Cache memory

Disadvantages:

- It is slower than SRAM
- It requires refresh operation which is not needed in SRAM
- Higher power consumption than SRAM due to continuous refreshing operation
- Data is not retained after removing power

5.4.4 Double Data Rate Synchronous Dynamic RAM (DDR SDRAM)

The DDR SDRAM is a type of RAM that is used in modern day processors. These are the memories that can transfer data at twice as fast as regular memories. The main advantage of DDR SDRAM is that it can fetch data at both "rising and falling edge" of the "clock signal", thus "double the data rate" for given "clock frequency". The types of DDR SDRAM are DDR1, DDR2, DDR3 and they word on supply voltage 2.5V, 1.8V, 1.5V respectively.

Advantages:

- Much faster speed than regular DRAM
- Each generation is updated like DDR2, DDR3, DDR4.

Disadvantages:

- It cannot be used in old motherboards
- Pinout is different from others
- Power consumption is high

5.5 READ ONLY MEMORY (ROM)

The ROM is a "semiconductor memory" that is used to "store the data/information permanently". It is non-volatile that means the "data is retained" in the "memory even after the supply is OFF". As the name indicates; it can perform only read operation or a user cannot change/alter its permanently stored date. It is designed for a particular purpose during manufacturing.

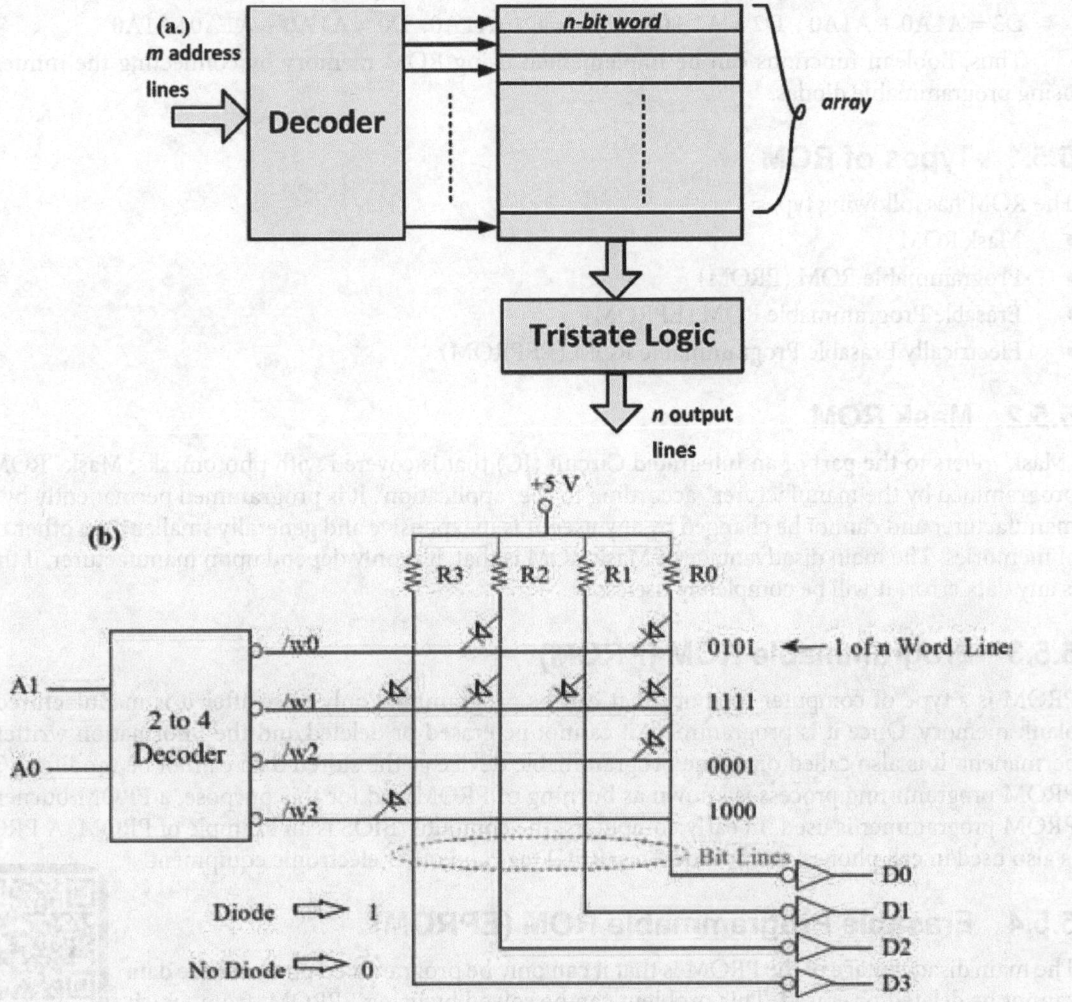

Fig. 5.11: (a) Block Diagram of ROM, (b) Internal structure of 4X4 ROM.

A ROM is basically a combination of AND and OR array that can programmed by a user to implement Boolean functions. The general "organization/block diagram" of "ROM" is shown in Fig. 5.11(a) and Fig. 5.11(b) represents internal structure of 4X4 ROM. As shown in Fig. 5.11(a), ROM has "m address lines" for decoder (decoder gives minterms); that means "2m memory locations" and "n output lines". Each bit combination of address variables is known as "address". The ROM is characterized by the number of "words 2m" and the number of "bits per word n". The size of the ROM is defined as 2m x n; that means the ROM consists of 2m words of n bit each.

Fig. 5.11(b) shows a 4X4 ROM here m=2 and n=4. Therefore, a 2X4 decoder is used to create 2m =4 address lines and each word is having n=4 bits. So, the total storage capacity of this ROM is 4 words or 4X4=16 bits.

AND plane is the decoder while OR plane is created using diode connections for whatever set of minterms is TRUE, we want to include for that bit position to be TRUE. Here D0 to D3 are representing OR operations for the minterms which are true. Thus, these can be represented as-

$D3 = A1'A0 + A1A0$; $D2 = A1'A0' + A1'A0$; $D1 = A1'A0$; $D0 = A1'A0' + A1'A0 + A1A0'$

Thus, Boolean functions can be implemented using ROM memory by connecting the minterms using programmable diodes.

5.5.1 Types of ROM

The ROM has following types:

- Mask ROM
- Programmable ROM (PROM)
- Erasable Programmable ROM (EPROM)
- Electrically Erasable Programmable ROM (EEPROM)

5.5.2 Mask ROM

"Mask" refers to the part of an Integrated Circuit (IC) that is covered with photomasks. Mask "ROM is programmed by the manufacturer" according to the "application". It is programmed permanently by the manufacturer and cannot be changed by any user. It is inexpensive and generally smaller than other type of memories. The main disadvantage of Mask ROM is that, it is only depend upon manufacturer, if there is any data error; it will be completely useless.

5.5.3 Programmable ROM (PROM)

PROM is a type of computer memory that can be programmed only once after it is manufactured as blank memory. Once it is programmed, it cannot be erased or deleted and the information written is permanent. It is also called one-time programmable device as the stored data cannot be modified. The PROM programming process is known as burning of PROM and for this purpose, a PROM burner or PROM programmer is used. In early computers, the computer BIOS is an example of PROM. A PROM is also used in cell phones, medical devices, RFID tags and other electronic equipment.

5.5.4 Erasable Programmable ROM (EPROM)

Secondary Memory

The main disadvantage of the PROM is that it can only be programmed once and the data cannot be deleted or erased. This problem can be solved by using EPROM, from which data can be erased by placing it under special ultraviolet light for some time. The EPROM returns to initial state by shortwave radiation and then it can be programmed again. It is used in some microcontrollers like Intel 8048.

5.5.5 Electrically Erasable Programmable ROM (EEPROM)

EEPROM is similar to EPROM, but in this the data is erased by applying electrical signal instead of ultraviolet light. It provides ease in erasing because it can be done even if the memory is inside the computer. It can erase or write only one byte of data at a time. It is used in computer BIOS.

5.5.6 Flash Memory (Flash ROM)

Flash memory is the enhanced version of EEPROM. In flash memory, blocks of data can be erased or written at a time, whereas, in EEPROM, only one byte of data can be erased. Thus, flash memory is much faster than EEPROM.

All the modern-day computers have their BIOS stored in flash memory and it is also used in MODEMs.

5.6 SECONDARY MEMORY

The memory devices which are connected to a computer externally or built into computer are called secondary memory/external memory/auxiliary memory. It can store data permanently "even when the system is turned OFF", so it is "non-volatile memory". It cannot be accessed directly by the computer; firstly, the data of secondary memory is transferred to primary memory and then CPU can access it. Some types of secondary memory are given below-

5.6.1 Hard Disk

A "hard disk" is a "magnetic disk" used to "store data permanently". It is located in the drive unit of motherboard of computer. The examples of stored data are installed software, images, songs, music videos, documents etc.

5.6.2 Solid-State Drive

"Solid-state drive (SSD)" is another type of "storage device". It has "faster access time" and "lower power consumption" than hard disk. It is ideal replacement for hard disk due to less cost. It is mostly used in modern laptops, tablets.

5.6.3 Pen Drive

A pen drive or USB flash drive is a compact secondary storage device which connects via a USB port of a computer. The data can be transferred to any other computer using pen drive. It is very small in size with storage capacity range from 1GB to 64GB or more.

5.6.4 Secure Digital Card (SD Card)

A Secure Digital Card also known as SD Card, is mostly used in portable or mobile devices including smartphone, digital cameras and other smart electronic devices.

Apart from above examples, some other types of secondary memory are:

- Compact Disk (CD)
- Digital Versatile Disc (DVD)
- Magnetic Tape
- Floppy Disk
- Blue Ray Disk

Cache Memory

5.7 CACHE MEMORY

Cache memory is a part of primary memory that stores the data or information which is frequently used so that it can be accessed by processor in less time. A CPU always checks the cache memory first to access the data, if it is not found in cache then the processor checks the main memory. The cache memory is placed between the primary memory and CPU. The cache memory is much faster than primary memory (RAM). The block diagram is shown below in Fig. 5.12.

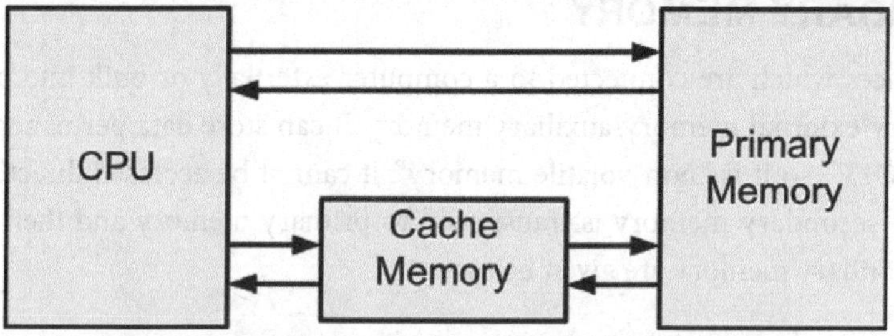

Fig. 5.12: A general placement of cache memory.

Cache Hit: If the data is found in the cache memory, it is said to be cache hit. The ratio of cache hits to total lookups is known as cache hit ratio.

$$\text{cache hit ratio (\%)} = \frac{\text{number of cachehits}}{(\text{number of cachehits + number of cachemiss})} \times 100\%$$

Cache Miss- If the data is not found in the cache memory, it is said to be cache miss.

Since SRAM is bigger in size (6T) thus the cache size is kept with low capacity (Kbytes) and sometimes secondary cache is used to make the processor faster. Primary cache has more frequently used data while secondary cache has relatively less frequently used data. First, the processor searches in primary cache and if the primary cache gives cache miss then it searches data in secondary cache. If the data is not found in secondary cache then it searches in main memory. Since main memory is bigger in size, so it takes more time to search the data in main memory as compared to the cache memory. Secondary cache is placed between primary cache and main memory.

5.8 DIFFERENCE BETWEEN RAM AND ROM

The basic difference between RAM and ROM is shown in Table 5.1.

Table 5.1: Comparison of RAM and ROM.

RAM	ROM
Volatile Memory	Non-volatile memory
Data can be read and altered	Data can only be read
Speed is high	Speed is slower than RAM
The CPU can directly access the data stored in RAM	The CPU cannot directly access the data stored in ROM, it needs to be stored in RAM
Size is small	Size is large
It can store only a few Megabytes of data	It can store Gigabytes of data
It is used on CPU, example Cache memory uses SRAM	It is used in micro-controllers

Example 5.1

A memory has a capacity of 6K x 8. Calculate-

(a) The number of data input and output lines of memory

(b) The number of address lines

(c) The capacity in bytes

Solution:

(a) The number of data input and output lines=8

(b) The number of address lines

Since Total addresses of memory 6K=6 x 1024=6144<2^{13}

Thus 13 address lines are required.

(c) 8 bits =1 byte; the capacity is bytes=6144 bytes

Example 5.2

Build a memory with storage of 2^{24} bits with 4 byte words. Calculate-

(a) The number of 2K × 8 bits RAM chips required

(b) The number of address lines required for memory

Solution:

(a) Given that the RAM capacity=2K × 8 bits=2 × 2^{10} × 2^3 bits = 2^{14} bits

Required memory capacity = 2^{24} bits

So, the number of RAM chips required = $2^{24}/2^{14} = 2^{10}$ = 1024

(b) Given that, a word is of 4 bytes

The number of possible words= $2^{24}/(4 \times 8) = 2^{24}/32 = 2^{24}/2^5 = 2^{19}$

So, the number of address lines=19

5.9 IMPLEMENTATION OF BOOLEAN LOGIC USING ROM

ROM can also be used to implement combinational logic. See the example below-

Example 5.3

Implement the following function using ROM- F1 (A, B, C) = Σm (0, 2, 4) and F2 (A, B, C) = Σm (1, 2, 5, 7).

Solution:

Truth Table:

A	B	C	F_1	F_2
0	0	0	1	0
0	0	1	0	1
0	1	0	1	1
0	1	1	0	0
1	0	0	1	0
1	0	1	0	1
1	1	0	0	0
1	1	1	0	1

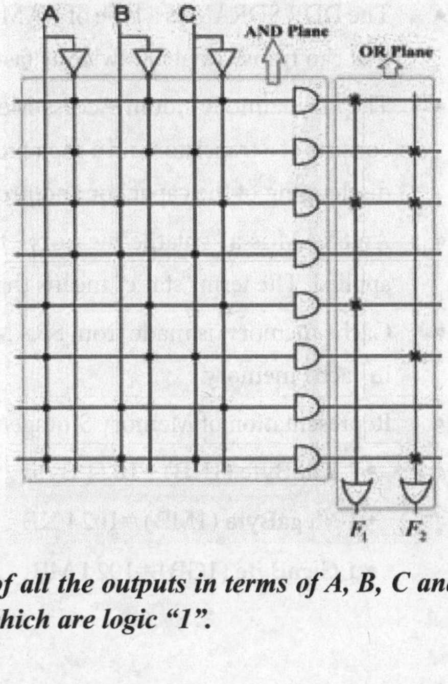

$F_1 = \bar{A}.\bar{B}.\bar{C} + \bar{A}.B.\bar{C} + A.\bar{B}.\bar{C}$

$F_2 = \bar{A}.\bar{B}.C + \bar{A}.B.\bar{C} + A.\bar{B}.C + A.B.C$

<u>Note</u>: The "dot (.)" represents the AND combination of all the outputs in terms of A, B, C and the "cross (x)" represents the OR combination of outputs which are logic "1".

UNIT SUMMARY

- A "memory" is an important "part of computer" and any electronic based system that is used to store data, program and instructions used for executing any program.
- The types of Memory -
  1. Primary Memory
     - "RAM"- "Static RAM (SRAM)", "Dynamic RAM (DRAM)"
     - "ROM" - PROM, EPROM, EEPROM
  2. Secondary Memory
  3. Cache Memory
  4. Register Memory
- "Memory" is a "collection of storage cells", which stores data & information in group of bits knows as words. A word is a sequence of 0s and 1s and each word is stored in a memory location, known as address. The group of eight bits is knows as a byte and the total number of bytes stored by memory is known as capacity of memory or memory size.
- "RAM is volatile memory" that means it cannot store data permanently; it is lost when the computer is turned OFF.
- "ROM is non-volatile memory" that means the "data is retained" in the "memory even after the supply is OFF". ROM is AND plane followed by OR plane. ROM can be used to implement Boolean functions as well.
- The DDR SDRAM is a type of RAM that is used in modern day processors. These are the memories that can transfer data at twice as fast as regular memories.
- The "Dynamic Random Access Memory (DRAM)" is a type of "semiconductor memory" which consists of a transistor and a capacitor. It "stores each bit of data in a capacitor". It works on charging/discharging of the capacitor and provides states logic '1'/logic '0' accordingly.
- An SRAM is a "volatile memory" that means the "data is stored as long as the power supply" is applied. The term "static" means that it does not need to be refreshed for updating the data.
- Cache memory is made from SRAM cells. Cache hit if data found and cache miss is data not found in cache memory.
- Representation of Memory Storage/Capacity:
  - 1 KiloByte (1KB) =1024 Byte ; 1 Byte=8 bits ; 1 Nibble=4 bits
  - 1 MegaByte (1MB) =1024 KB ; 1 TeraByte (1 TB) = 1024 GB
  - 1 GigaByte (1GB) =1024 MB; 1 Peta Byte (1 PB) = 1024 TeraByte

EXERCISES

Multiple Choice Questions

1. For a memory, 1MB equals to-

(a) 100KB (b) 1024KB
(c) 512KB (d) 2KB

2. Which is not a secondary memory-

(a) SRAM (b) Had Disk
(c) Pen Drive (d) CD

3. Which is the smallest in size?

(a) KB (b) MB
(c) GB (d) TB

4. Identify the wrong statement-

(a) 1GB=1024MB (b) 1MB=1024KB
(c) 1KB=1024bytes (d) 1GB=1024KB

5. Which is a permanent memory?

(a) SRAM (b) DRAM
(c) Bipolar RAM (d) ROM

6. A DRAM consists of one ________ and one capacitor-

(a) Diode (b) Resistor
(c) Transistor (d) Inductor

7. ________ requires periodic refresh operation-

(a) SRAM (b) ROM
(c) DRAM (d) All

8. Which is enhanced version of EEPROM?

(a) MASK ROM (b) PROM
(c) Flash Memory (d) ROM

9. A memory has 5 address lines, how many numbers of words it can store?

(a) 16 (b) 32
(c) 5 (d) 8

10. A RAM is also known as________ memory-

(a) Read-Write (b) Read
(c) Write (d) Permanent

11. Memory is a collection of__________-

(a) Storage cells (b) Transistors
(c) Gates (d) Diodes

12. In RAM, bit-addressable area has__________ addresses-

(a) 64 (b) 128
(c) 256 (d) 32

13. For a memory with size 2K x 8, the number of address lines is-
   (a) 1024
   (b) 2048
   (c) 512
   (d) 4096

14. Which is used to identify storage?
   (a) Address
   (b) Array
   (c) Record
   (d) Bytes

15. For a memory, 2GB can also be represented as-
   (a) 2048MB
   (b) 1024MB
   (c) 100MB
   (d) 1000KB

16. Which is not a part of any memory unit?
   (a) Address lines
   (b) Read port
   (c) Write port
   (d) Reset

ANSWERS															
1.	(b)	2.	(a)	3.	(a)	4.	(d)	5.	(d)	6.	(b)	7.	(b)	8.	(c)
9.	(b)	10.	(a)	11.	(a)	12.	(d)	13.	(b)	14.	(d)	15.	(a)	16.	(d)

Short and Long Answer Type Questions

Category 1

1. Draw the basic structure of memory representing the address lines, read/write ports and input/output lines.
2. What is the difference between volatile and non-volatile memory?
3. What is Flash memory? How it is better than other ROMs?
4. Enlist types of register banks in RAM and Draw the structure.
5. Write the difference between EPROM and EEPROM.
6. Draw the general structure of CMOS based SRAM.
7. The size of a memory is 6K x 8, calculate the number of input/output lines and address lines.
8. What is cache hit ratio?
9. What is the function of OR plane in ROM?

Category 2

1. Explain the operation of CMOS based SRAM with suitable diagram.
2. A half adder has input U and V, the Sum and Carryout. Implement the structure using ROM.
3. Draw the general organization of memory and describe every ports associated with it.
4. Describe various types of ROM and write application of each of them.
5. Design a 3-bit binary code to 3-bit gray code converter using ROM.
6. Define the terms (a) memory address (b) memory cell (c) memory word (d) bits and byte (e) chip select (f) memory refresh
7. Draw the structure of 1T -1C -DRAM cell and write its advantages and disadvantages over 6T-SRAM cell.

8. Design a combinational circuit using ROM which has "3-bit input" and "output is equal to" the "2's complement of the input".
9. Design a combinational circuit using ROM which has "2-bit input" and "output is square of input". (Hint- input is 2-bit then square output is 4-bit, draw truth table and find expressions for each output)

Numerical Problems

1. A memory has a capacity of 8K x 12. Calculate-
   (a) The "number of data input" and "output lines" of memory
   (b) The "number of address lines"
   (c) The capacity in bits and bytes
2. Determine the type of decoder required, if a memory with a capacity of 223 bits (with 4 byte words) is built using 2K x 8 RAM.
3. What will be the bit, and byte storage capacity for a 512 x 8 ROM organization?
4. A certain memory can store 16K, 16-bit words. Calculate (a) the "number of input" and "output lines" (b) the "number of address lines" (c) capacity in bytes
5. A computer searches 1000 words in cache memory. It finds 199 words in cache. Find cache hit ratio.

PRACTICAL

EXPERIMENT 9. Study different Memory ICs.

SRAM IC

(1) LY62256 (LY is manufacturer name-Lyontek Inc., 62256 model number)

- Size=256Kbit
- DIP
- Fast access time of 55ns
- Ultra low power consumption
- Operating current of 15mA
- Standby current : 1μA
- Operating temperature range from -40°C to 85°C

The LY62256 is a (32KX8) 262144-bit low power CMOS static random access memory organized as 32,768 (32KX8) words by 8 bits. The LY62256 is well designed for low power application, and particularly well suited for battery back-up nonvolatile memory application. The LY62256 operates from a single power supply of 2.7~5.5V and all inputs and outputs are fully TTL compatible. It can also be written as LY62256PL-55LLI

P:28 pin 600mil P-DIP (plastic-DIP), L: Green Package, 55: access time

LL-ultra low power, I- (industrial): -40°C to 85°C

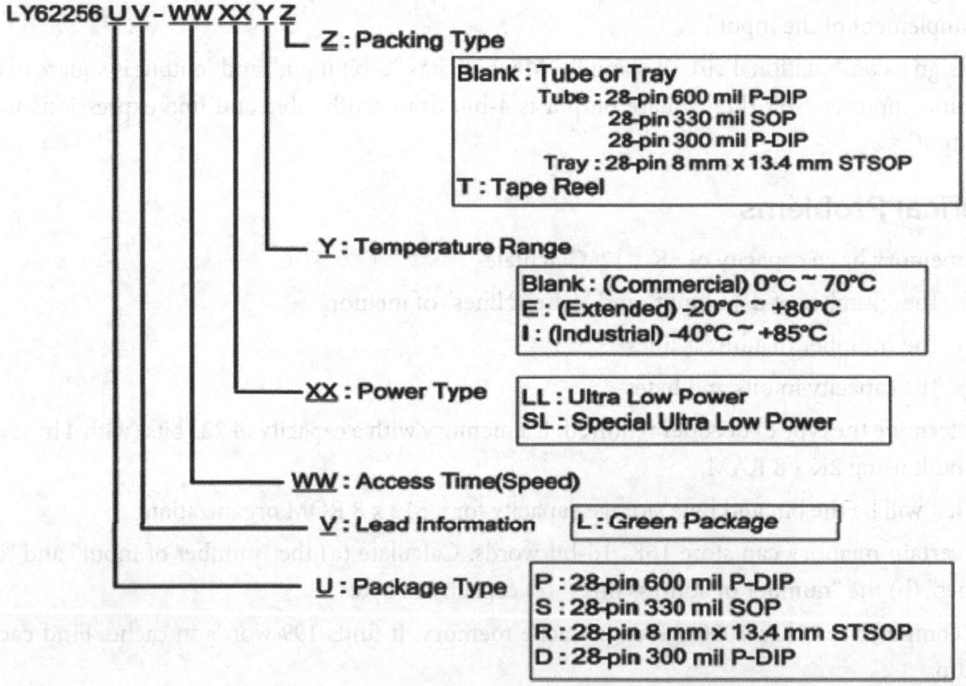

Pin Configuration

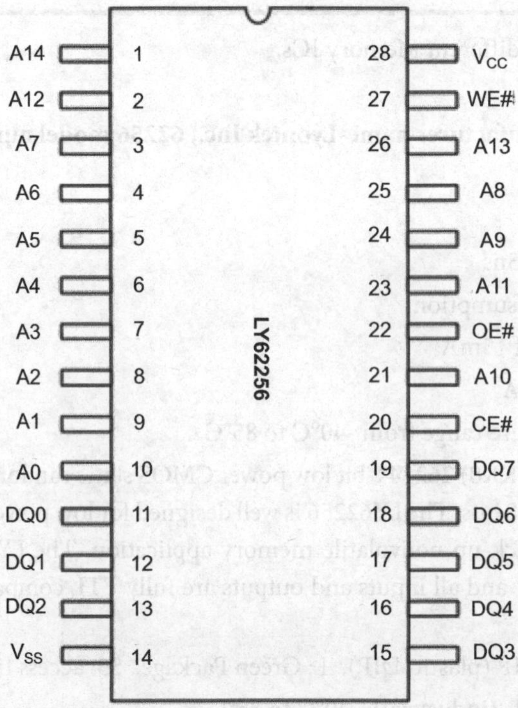

FUNCTIONAL BLOCK DIAGRAM

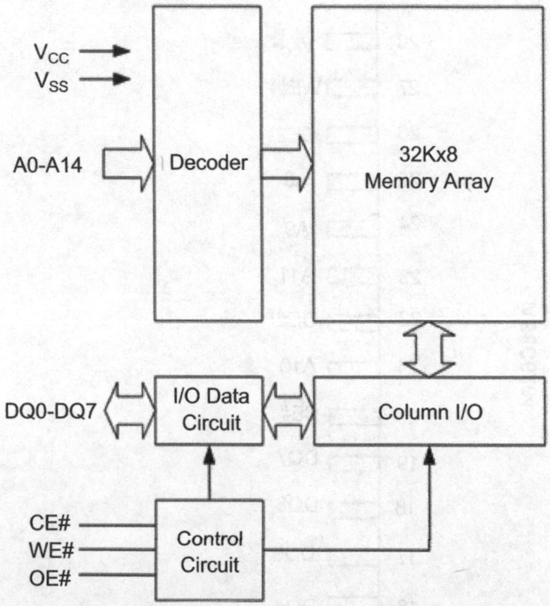

SYMBOL	DESCRIPTION
A0 - A14	Address Inputs
DQ0 – DQ7	Data Inputs/Outputs
CE#	Chip Enable Input
WE#	Write Enable Input
OE#	Output Enable Input
Vcc	Power Supply
Vss	Ground

Data sheet: https://www.farnell.com/datasheets/1674430.pdf

SRAM IC

(2) AS6C6264

- 8K X 8 BIT LOW POWER CMOS SRAM
- Access times: 55 ns
- Operation current : 15mA , VCC = 3.0V Standby current : 1μA , VCC = 3.0V
- Wide range power supply : 2.7 ~ 5.5V
- Fully static operation
- Tri-state output

The AS6C6264 is a 65,536-bit low power CMOS static random access memory organized as 8,192 words by 8 bits. It is fabricated using very high performance, high reliability CMOS technology. The AS6C6264 is well designed for low power application, and particularly well suited for battery back-up nonvolatile memory application

Pin Configuration

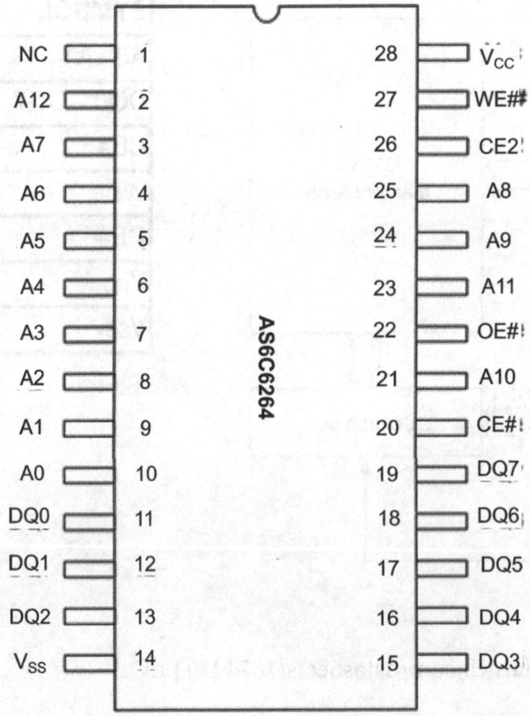

FUNCTIONAL BLOCK DIAGRAM

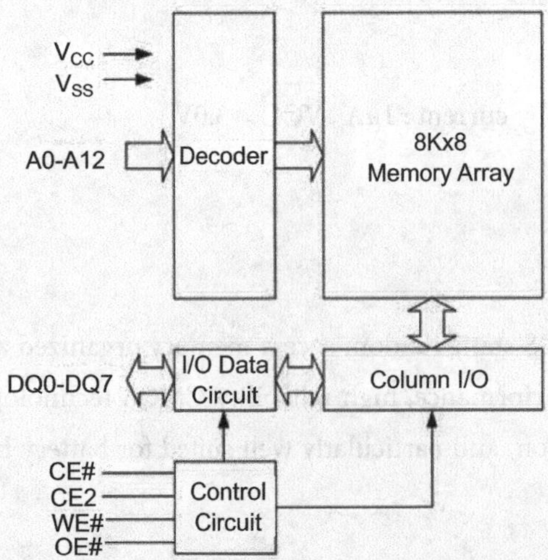

PIN DESCRIPTION

SYMBOL	DESCRIPTION
A0 - A12	Address Inputs
DQ0 – DQ7	Data Inputs/Outputs
CE#, CE2	Chip Enable Inputs
WE#	Write Enable Input
OE#	Output Enable Input
Vcc	Power Supply
Vss	Ground
NC	No Connection

Ordering Codes

Alliance	Organization	VCC range	Package	Operating Temp	Speed ns
AS6C6264-55PCN	8k x 8	2.7-5.5V	28pin 600mil PDIP	Commercial ~ 0° C to 70° C	55
AS6C6264-55SCN	8k x 8	2.7-5.5V	28pin 330mil SOP	Commercial ~ 0° C to 70° C	55
AS6C6264-55SIN	8k x 8	2.7-5.5V	28pin 330mil SOP	Industrial ~ -40°C to 85° C	55
AS6C6264-55STCN	8k x 8	2.7-5.5V	28pin sTSOP (8 x 13.4 mm)	Commercial ~ 0° C to 70° C	55
AS6C6264-55STIN	8k x 8	2.7-5.5V	28pin sTSOP (8 x 13.4 mm)	Industrial ~ -40°C to 85° C	55

Part numbering system

AS6C	6264	- 55	X	X	N
low power SRAM prefix	Device Number 6264	Access Time	Package Options: P = 28 pin 600 mil P-DIP S = 28 pin 330 mil SOP ST = 28 pin sTSOP (8mm x 13.4 mm)	Temperature Range: C = Commercial (0°C to +70° C) I = Industrial (-40° to +85° C)	N = Lead Free ROHS Compliant Part

Data Sheet : https://docs.rs-online.com/64ba/0900766b80b7b917.pdf

Other examples: https://in.element14.com/c/semiconductors-ics/memory/sram

DRAM

MB8116 (MOS 16384-bit DRAM)

The Fujitsu MB8116 is a fully dynamic NMOS DRAM organized as 16384 one-bit words.

Features:

- 16384x1 RAM
- 16 pin DIP
- Silicon gate, single transistor cell, double poly NMOS
- Low power 462mW (active), 20mW (stand by)
- Cycle time: 375ns (minimum)
- All inputs TTL compatible
- Three state TTL Compatible output
- 128 refresh cycles
- Operating Temperature: 0 to 70 degree

Pin Assignment

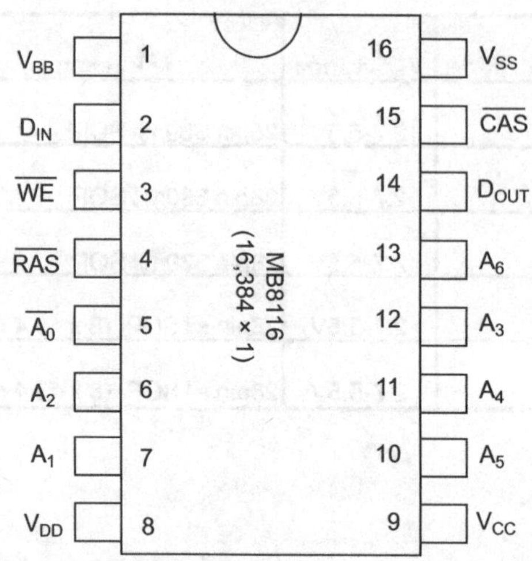

Data sheet: https://www.jameco.com/Jameco/Products/ProdDS/41339%20Fujitsu.pdf

KNOW MORE

Over the past 90 years, data storage evolved from magnetic drums and tapes to hard disk drives, then to mixed media, flash, and finally cloud storage. That's where we are today, and as our storage needs increase, innovation continues to evolve in multiple areas. Cloud option is good to handle big data and characterise it at different ends all together. Big data is needed since now we are in 5G where data is coming for smart vehicles, smart sensors others. Preprocessing data can give us many timely alarms. For example- data from space satellites is characterised to know exact weather conditions in the country. Data from space satellites can give pre-calculated production of crops at different regions of the country.

REFERENCES AND SUGGESTED READINGS

— Digital Design by M. Moris Mano and Michael D Ciletti, 5th edition, Pearson

— Digital Electronic Circuits - Course (nptel.ac.in)

Dynamic QR Code for Further Reading

QR codes embedded in the chapter

□□

6 Data Converters

UNIT SPECIFICS

Through this unit we have discussed the following aspects:

- Digital To Analog Converters(DAC): Types and Specifications
- *Sample and Hold (S/H) Circuit;*
- Analog-To-Digital Converters (ADC): Quantization and Encoding;
- Specifications and Types of ADC Converters;
- Example of ADC & DAC Converter ICs

The practical applications of the topics are discussed for generating further curiosity and creativity as well as improving problem solving capacity.

Besides giving a large number of multiple choice questions as well as questions of short and long answer types marked in two categories following lower and higher order of Bloom's taxonomy, assignments through a number of numerical problems, a list of references and suggested readings are given in the unit so that one can go through them for practice. It is important to note that for getting more information on various topics of interest some QR codes have been provided in different sections which can be scanned for relevant supportive knowledge.

After the related practical, based on the content, there is a "Know More" section. This section has been carefully designed so that the supplementary information provided in this part becomes beneficial for the users of the book. This section mainly highlights the initial activity, examples of some interesting facts, analogy, history of the development of the subject focusing the salient observations and finding, timelines starting from the development of the concerned topics up to the recent time, applications of the subject matter for our day-to-day real life or/and industrial applications on variety of aspects, case study related to environmental, sustainability, social and ethical issues whichever applicable, and finally inquisitiveness and curiosity topics of the unit.

RATIONALE

This fundamental unit on converters helps students to get a primary idea about the digital to analogue to digital (ADC) converters. A DAC converter is crucial because it is required at the output of the majority of digital systems, where it converts a digital signal into analogue voltage or current so that it can be fed to a chart recorder, for example, for measurement purpose, or a servo motor for a control application. While ADC convertering analogue data, where the analogue signal to be communicated is converted to digital at the sending end using an ADC converter. This chapter discusses operational basics, the key performance criteria and their significance, as well as the many forms and uses of digital-to-analogue and analouge-to-digital converters. Solved examples are used to effectively explain the chapter.

PRE-REQUISITES

Boolean algebra, Combinational & Sequential circuits, Number system, memory

UNIT OUTCOMES

List of outcomes of this unit is as follows:

U6-O1: Describe DAC

U6-O2: Describe S/H Circuit;

U6-O3: Describe ADC

U6-O4: Specifications and Types of Converters

U6-O5: Example of ADC Converter IC

Unit-6 Outcomes	EXPECTED MAPPING WITH COURSE OUTCOMES (1 - Weak Correlation; 2 - Medium Correlation; 3 - Strong Correlation)					
	CO-1	CO-2	CO-3	CO-4	CO-5	CO-6
U6-O1	2	2	3	-	-	3
U6-O2	-	-	-	-	2	3
U6-O3	2	2	3	-	-	3
U6-O4	-	-	-	-	-	3
U6-O5	-	-	-	-	-	3

"Knowledge will bring you the opportunity to make a difference."

—CLAIRE FAGIN

6.1 INTRODUCTION

Converters from digital to analog (DAC) and from analog to digital (ADC) are a necessary component to interface between digital and analog devices. In addition to having various applications, they are crucial components of any system, specially the systems which has digital processing. A DAC converter is crucial for digitally processed signals to be converted to analog for machine human interface, for example, speaker, or a servo motor for a control application. A DAC converter is also crucial because it is a crucial component of the majority of ADC converter types. ADC converters have a wide range of uses as well. All digital read-out test and measurement equipment uses it without fail. An ADC converter is a crucial and integral part of any digital multimeter, as well as digital storage oscilloscopes, pH metres etc. The operational basics, the key performance criteria and their significance, as well as the many forms and uses of DACs and ADCs will all be covered in this UNIT.

Based on the conversion speed, DAC and ADC are of two types:

1. **Nyquist rate converters:** If the speed of conversion is 1.5 to 10 times of its bandwidth or 3 to 10 times of baseband frequency, ADC or DAC will be called as Nyquist rate converter.
2. **Oversampling converters:** If the speed of conversion is more than Nyquist rate converters, ADC or DAC is known as oversampling converter.

6.2 DIGITAL-TO-ANALOG CONVERTERS (DAC)

When digital data is input (b_0, b_1, b_2 etc.) a DAC converter translates it into analog voltage or current that is proportionate to the weighted sum of the input bits b_0, b_1, b_2, etc. There are various types of DAC as discussed in the subsequent subsections.

6.2.1 Weighted Resistor DAC

A weighted register or binary weighted resister DAC generates an analog output that is nearly equivalent to the digital input by utilising binary weighted resistors at the inverting node of OPAMP based adder circuit as shown in Fig. 6.1.

Remember that a binary bits can only have one of two possible values that is, either 0 or 1. Allow $b_2b_1b_0$ to be the 3-bit binary input. The Most Significant Bit (MSB) and Least Significant Bit (LSB) are indicated here by the bits b_2 and b_0, respectively. When the respective input bits equal "0", the digital switches in the above diagram will be connected to the ground. Similar to this, when the relevant input bits are equal to "1", the digital switches illustrated in the image will be linked to the negative reference voltage, VR. The non-inverting input terminal of an OPAMP is linked to the ground in the circuit shown in Fig. 6.1. That indicates that "0" volts are applied to the non-inverting input terminal of OPAMP which in turn makes the virtual ground at the inverting terminal. Thus, Kirchhoff's Current Law at inverting node of OPAMP can be written as:

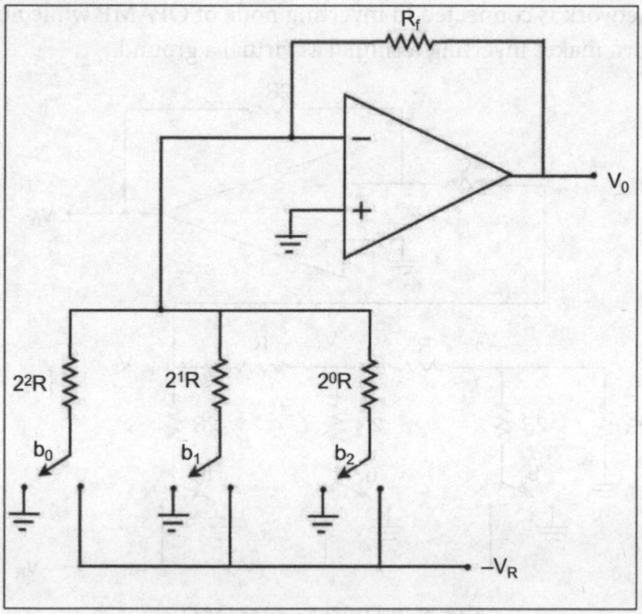

Fig. 6.1: Weighted Register DAC.

$$\frac{0+V_R b_2}{2^0 R}+\frac{0+V_R b_1}{2^1 R}+\frac{0+V_R b_0}{2^2 R}+\frac{0-V_0}{R_f}=0$$

$$\Rightarrow \quad \frac{V_0}{R_f}=\frac{V_R b_2}{2^0 R}+\frac{V_R b_1}{2^1 R}+\frac{V_R b_0}{2^2 R}$$

$$\Rightarrow \quad V_0=\frac{V_R R_f}{R}\left\{\frac{b_2}{0^0}+\frac{b_1}{2^1}+\frac{b_0}{2^2}\right\}$$

This equation represents the analog output V0 for a 3-bit binary weighted resistor DAC's output. The binary input can take value from 000 to 111 and reference voltage VR is fixed. The above equation can be generalised for an N-bit binary DAC as -.

$$\Rightarrow \quad V_0=\frac{V_R}{2}\left\{\frac{b_{N-1}}{2^0}+\frac{b_{N-2}}{2^1}+....+\frac{b_0}{2^{N-1}}\right\}$$

Although this DAC is easy to design but there are some drawbacks of this DAC such as:

1. There are N different resistors for N-bit DAC which are multiples of 2i for ith bit.
2. Since each resistor is different, so it is very tough to maintain accuracy in multiples of 2i for ith bit because tolerance parameter affects the value of any resistor.

6.2.2 R-2R Ladder DAC Converter

This ADC uses R-2R Ladder network instead of N-different resistors. The R-2R Ladder network uses just two resistors R and just double of that 2R for N-bit DAC. The DAC circuit diagram is shown in Fig. 6.2 where the resistor network is connected to inverting node of OPAMP while non-inverting terminal is grounded which in turn makes inverting terminal as virtual a ground.

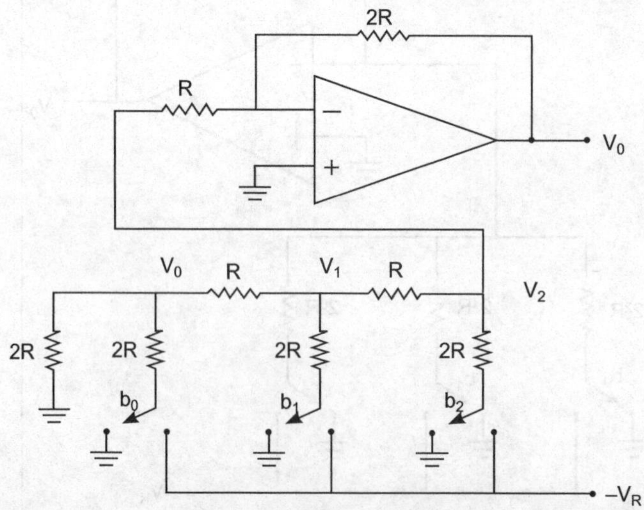

Fig. 6.2: R-2R Ladder DAC

Remember that each of the two possible values for a binary number's bits can only occur once. It's either LOW or HIGH. Assume $b_2b_1b_0$ as binary input with b2 and b0 presented as MSB & LSB respectively. In the example above, when the matching input bits equal "0", the digital switches are connected to ground and for any bit is "1" that switch is connected to "VR". It is simple to determine the R-2R Ladder DAC's analog output voltage values for specific binary input combinations. From the below Fig. 6.2, it is clear that the current in b0 branch is minimum (let's say I0) and for every higher ith bit, this current is 2i.I0. This can be easily derived using KCL at every node. Thus, it gives the result corresponding to the weighted resistor type DAC only.

A R-2R Ladder DAC has the following benefits:

1. There are only two resistor values included - R and 2R. So choosing and creating more precise resistors is simple.
2. If the digital input contains more bits, then we must also include the necessary number of R-2R sections.

Example 1: $b_2b_1b_0 = 100$:- Here since $b_1b_0 = 00$, therefore total resistance seen at V_0 = 2R||2R = R and resistance seen at V1= (R+R)||2R = R. Since $b_2 = 1$, so it is connected to –VR. So, V2 = 2R/(2R+R+R)*(–VR) = (–VR)/2 which is input voltage for inverting terminal. Now using inverting terminal OPAMP formula -

$$V_0 = -\frac{2R}{R}V_2 = V_R$$

Example 2. $b_2b_1b_0 = 110$:- Here, more than one bit is HIGH, so superposition theorem is to be used to calculate the net V0.

Case 1: Assume b2b1b0=100: in this case the partial output is same as obtained in the previous example. Therefore due to MSB = 1, output is

$$V'_0 = -V_R$$

Case 2: Assume b2b1b0=010: Here since b0=0, therefore total resistance seen at V0 = 2R||2R = R. Take the R-2R ladder out and solve to find V2.

Applying nodal analysis at V1-

$$\frac{V''_1}{R+R} + \frac{V''_1 + V_R}{2R} + \frac{V''_1}{3R} = 0$$

$$\Rightarrow V_1''\left(\frac{1}{2R} + \frac{1}{2R} + \frac{1}{3R}\right) = -\frac{V_R}{2R}$$

$$\Rightarrow V''_1 = \frac{-V_R/2}{\left(\frac{1}{2} + \frac{1}{2} + \frac{1}{3}\right)} = \frac{3}{2}\left(\frac{-V_R}{4}\right)$$

From voltage division rule $V''_2 = \frac{2R * V_1''}{3R} = \frac{2}{3}\left(\frac{3}{2}\left(\frac{-V_R}{4}\right)\right)$ or $V''_2 = \left(\frac{-V_R}{4}\right)$

Now using inverting mode OPAMP formula output due to b_1 bit is given as

$$V''_0 = -\frac{2R}{R}V_2 = 2\left(\frac{V_R}{4}\right) = \left(\frac{V_R}{2}\right)$$

Thus the total output can be obtained by using the superposition theorem-

$$V_0 = V_0' + V_0'' = 1.5\ V_R$$

Note: It can be noted that the V_2 nodal voltage is $-\frac{V_R}{2}$ for MSB = 1 and keeps on reducing by a factor of two for subsequent bits. Likewise it can be obtained for $b_0 = 1$ case that $V_2 = -\frac{V_R}{8}$ and $V_0''' = \frac{V_R}{4}$.

Thus for "111" inputs $V_0 = \frac{V_R}{4} + \frac{V_R}{2} + \frac{V_R}{1} = 1.75\ V_R$.

6.3 DAC CONVERTER SPECIFICATIONS.

Resolution, precision, dynamic range, conversion speed, integrated nonlinearity (INL) and differential nonlinearity (DNL), and monotonicity are among the main performance parameters of a DAC converter.

6.3.1 Resolution

In an n-bit DAC's resolution is the quantity of all states (2^N) into which the full-scale voltage (V_{ref}) or current range is divided. The resolution improves with increasing bit count. There are 256 resolvable levels in an 8-bit DAC. It is claimed to have an eight-bit resolution, or (1/256)*100 = 0.39%, as its percentage resolution. A 12-bit DAC's percentage resolution would be (1/4096)*100 = 0.0244%. Typically, the resolution of an N-bit DAC is given in Volts -

$$V_{LSB} = {V_{ref}}/{2^N}$$

For the two scenarios, the resolution in millivolts for $V_{ref} = 10$ V is roughly 40 mV & 2.4 mV for 8 & 12-bit DAC respectively. 1 LSB is defined as 1-digital level for DAC which is a unitless quantity and mathematically can be represented as (sometimes resolution is defined as 1 LSB)

$$1LSB = \frac{V_{LSB}}{V_{ref}} = \frac{1}{2^N}$$

6.3.2 Accuracy

Accuracy is decided by deviation in actual output from the ideal one. Gain error (also known as full-scale error), offset error (also known as zero-scale error), nonlinearity errors, and a drift of all components are examples of error sources. The gain & offset errors are presented in Fig. 6.3, these are generally measured in LSB units. Foe example for $V_{ref} = 5$ V, an accuracy of 0.1% would imply that there can be maximum 5 mV deviation in the actual and ideal analog output.

6.3.3 Settling Time

When there is a change in a digital input, the analog output must have had enough time to settle within the permissible error band before its final value. This is known as the settling time. Several microseconds is the settling time of general-purpose DAC converters, while a few nanoseconds is the settling time of high-speed DAC converters. For instance, Analog Devices USA's AD 9768 type of DAC converter has 5 ns settling time criteria.

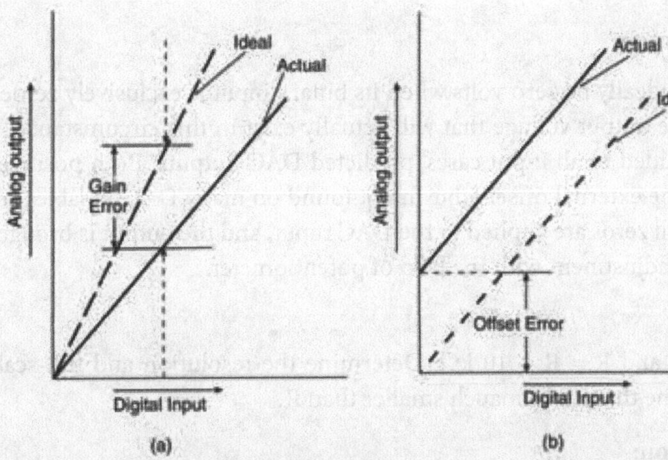

Fig. 6.3: (a) Gain error and (b) offset error.

6.3.4 Dynamic Range

In terms of decibels, this is the ratio of the highest output sinusoidal signal to the lowest noise plus distortion present at the output which can at least produce 1LSB. It is 20*log10(2N) for linear DAC converters, which is roughly equivalent to 6N decibel, where n- is the number of bits in DAC. It is commonly 66 or 72 dB for companding-type DACs.

6.3.4 Monotonicity

A perfect DAC must increase the analog output for increasing the binary input word, as shown in Fig. 6.4. Monotonicity implies that the output of DAC will always have one sign for its Vin versus Vout slope. It ensures monotonicity if the DAC's DNL error ≤ (twice its worst-case INL) i.e. 1 LSB OR DAC must have INL lesser than equal to 0.5 LSB.

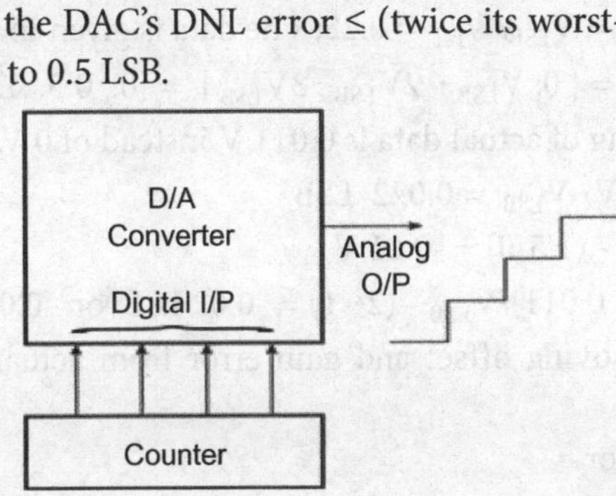

Fig. 6.4: Monotonicity in a DAC converter.

6.3.5 Integrated Nonlinearity and Differential Nonlinearity

Integrated nonlinearity (INL) is defined as deviation of analog output voltage from ideal straight line output. INL can be up to ±0.5 LSB for error free conversion.

The worst-case departure of any consecutive analog outputs from the ideal step size is known as differential nonlinearity (DNL). DNL can be ±1 LSB in worst case for error free conversion. DNL is the difference between consecutive INL errors.

6.3.6 Offset Error

DAC's output should ideally be zero volts when its binary input is exclusively zeroes. Offset error is the term for the very little output voltage that will actually exist in this circumstance. If this offset issue is not fixed, it will be added to all input cases' predicted DAC outputs. Both positive and negative offset errors are possible. The external offset adjustment found on many DACs enables you to zero the offset. Typically, to do this, all zeros are applied to the DAC input, and the output is brought as near as possible to "0" using an offset adjustment with the help of potentiometer.

Example 6.1

Assume V_{RFF} = 10 V and R = R = 10 kΩ. Determine the resolution and full-scale output for binary weighted DAC. Assume that RL in mauch smaller than R.

Answer & Explaination:

The $I_0 = V_{RFF}/R$ = 1 mA. This is the weight of the MSB. The other three currents will be 0.5, 0.25 and 0.125 mA. The LSB is 0.125 mA, which is also the resolution. The full-scale output will occur when the binary inputs are all HIGH so that each current swithc is closed and I_{OUT} = 1 + 0.5 + 0.25 + 0.125 = 1.875 mA. Note that the output current is proportional to V_{REF}. If V_{REF} is increased or decreased, the resolution and full-scale output will change proportionally.

Example 6.2

Consider a 2-bit DAC with V_{ref} = 2 V, with the measured values {0.011: 0.507 : 1.002 : 1.501}. Find-

Offset and gain error in LSB 2. INL and DNL errors (in LSB)

Solution:

Given that 2-bit DAC win V_{ref} = 2 V thus

$V_{LSB} = 2/2^2 = 0.5$ V; Unit less 1 LSB = V_{LSB}/V_{ref} = 0.25; The data is given in terms of voltage-

Ideal data set (in terms of voltage) = { 0: V_{LSB} : $2V_{LSB}$: $3V_{LSB}$ } = {0: 0.5 :1.0 :1.5}

1. **Offset error** : since the starting of actual data is 0.011 V instead of 0 V. Thus offset error = 11 mV.

   In terms of LSB offset = 11 mV/ V_{LSB} = 0.022 LSB

   Gainerror = (1.501 – 0.011) - (1.5-0) = -0.01 V

   Gain error in LSB = (1.501 – 0.011)/V_{LSB} -(2^2-1) = -0.02 LSB or –0.01 V/0.5 = –0.02 LSB
2. INL can be obtained by removing offset and gain error from actual data and then finding the difference from ideal data.

   Removing offset and gain error –

   For example data corresponding to 1.002 V can be obtained as –

1.01 - offset – gain*2 /(2^2–1) = 1.001 – 0.011 + 0.01*2/3 = 0.996 V or 1.993 LSB

Thus offset free and gain free values will be (in terms of voltage)

{0 : 0.499 : 0.996 :1.5 } or In terms of LSB {0: 0.9986 : 1.993 : 3}

thus **INL is (in LSB)** can be obtained from finding the difference of actual dat from the ideal one-

{0: 0.9986 -1 : 1.993 –1 : 3-3} = {0: – 0.0014: – 0.007 : 0}

DNL (in LSB) = {-0.0014: -0.0056 :0.007}

Example 6.3

A 5-bit DAC converter produces V_{OUT} = 0.2 V for a digital input of 0001. Find the value of V_{out} for an input of 11111.

Answer & Explanation

Obviously, 0.2 V is the weight of the LSB. Thus the weights of the other bits must be 0.4 V 0.8 V, 1.6 V and 3.2 V respectively. For a digital input of 11111, then the value of V_{OUT} will be 3.2 V + 1.6 V + 0.8 V + 0.4V – 0.2 V – 6.2 V.

Example 6.4

For the DAC of Example 2.3 determine V_{OUT} for a digital input of 10001.

Answer & Explanation

The step size is 0.2 V, which is the proportionality factor K. The digital input is 10001 – 1710. Thus we have : V_{OUT} – (0.2 V) × 17 – 3.4 V.

Example 6.5

A 10-bit DAC has a step size of 10 mV. Determine the full-scale output voltage and the percentage resolution.

Answer & Explanation

With 10-bits, there will be 2^{10} – 1 = 1023 steps of 10 mV each. The fullscale output will therefore be 10 mV × 10.23 V and % resolution –10 mV/10.23 V × 100% ≈ 0.1%.

For an N-bit binary input code the total number of steps is 2^N – 1. Thus, for the previous example, % resolution = 1/(210 – 1) × 10% ≈ 0.1%

Example 6.6

A certain 8-bit DAC has a full-scale output of 2mA and a full-scale error of ± 0.5% F.S. What is the range of possible outputs for an input of 10000000? A certain 8-bit DAC has a full-scale output of 2mA and a full-scale error of ± 0.5% F.S. What is the range of possible output for an input of 10000000?

Answer & Explanation

The setp size is 2 mA/255 = 7.84 μA. Since 10000000 = 12810, the ideal output should be 128 × 7.84 μA. THe error can be as much as ± 0.5% × 2mA = ± 10μA . Thus, the actual output can deviate by this amount from the ideal 1004 μA , so the actual can be anywhere from 994 to 1014 μA.

6.4 MUSEFUL CIRCUITS FOR ADC CONVERTERS

Electrical analog signals are converted to digital signals for data processing using analog to digital converters (ADC). The largest ADC converter portfolio in the market is provided by analog devices with solutions that meet requirements for performance, power, cost and size. Symbolic representation for ADC is shown in Fig. 6.5.

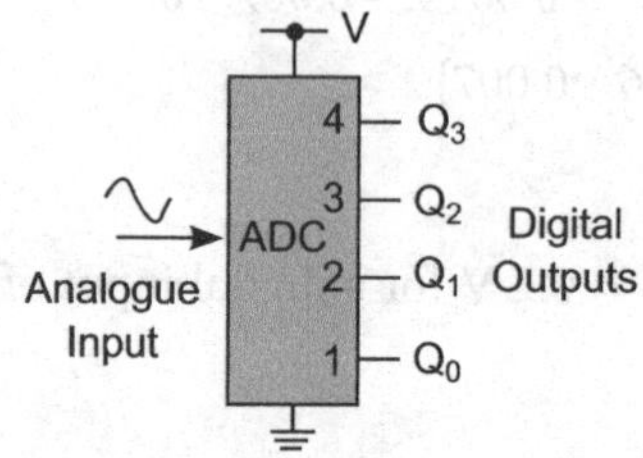

Fig. 6.5: Symbolic representation of analog to digital converter (ADC).

ADC converters, which are the best in the world, provide for precise and dependable conversion performance in a variety of applications including healthcare, energy, communications, instrumentation and measurement, industrial automation, motor and power control and aerospace/defence. The engineer can access a number of ADC converter resources to help at every stage of the project from component selection through circuit design.

Sample and Hold circuits, Quantization and encoding are very basic circuits used for ADCs.

6.4.1 Sample and Hold Circuit

The Sample and Hold (S/H) circuit is basic input circuit for ADC. S/H circuit creates voltage samples from input and then holds those samples for a set period of time. The sampling time refers to the period of time when the sample and hold circuit generates a sample of the input signal. Similar to this, holding time refers to how long the circuit keeps the sampled value in memory. The holding time can take on whatever value as necessary by the application, while the sampling time is typically between 1 and 14 seconds.

With the aid of an operational amplifier, the sample and hold circuit schematic is given in Fig. 6.6(b). The circuit schematic makes clear that a switch is used to link two OP-AMPS. The sampling process will start when the switch is closed, and the holding effect will start when the switch is opened and the data is hold on the capacitor connected at the input of another OPAMP.

Fig. 6.6: (a) Sample and Hold Circuit block diagram (b) Circuit diagram ofS/H Circuit using OP-AMP.

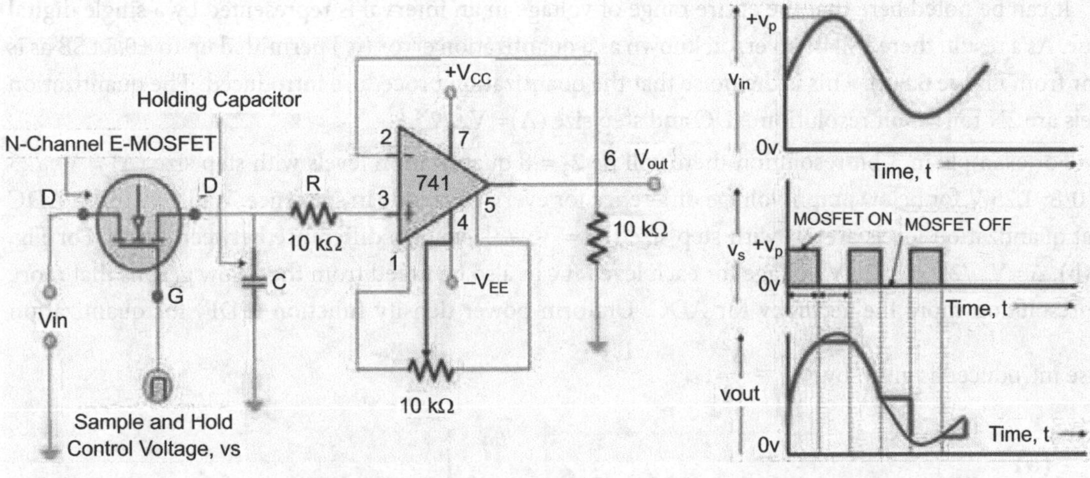

Fig. 6.7: Practical Circuit diagram and output ofS/H Circuit.

Working of Sample and Hold Circuit:

A simplest form of S/H circuit is a resistor –capacitor connected in series where the input is applied at the resistor and output is taken at the capacitor. The same circuit is presented in Fig. 6.6 in a modified format. The output is held constant during the OFF period of the clock.

A high precision unity gain OPAMP, a capacitor to store and retain the electric charge, and an N-channel enhancement type MOSFET are the primary parts of a practical S/H circuit as shown in Fig. 6.6.

As a switching component, the N-channel Enhancement MOSFET will be employed. Through its drain terminal, the analog input (AI) voltage is applied, and through its gate terminal, the control voltage (C) is applied. The MOSFET will be switched to the ON state when the C=1 is supplied and hold capacitor (CH) gets charged to its maximum through MOSFET. When C=0 the MOSFET is OFF which results to stops charging for capacitor CH. Since the input impedance of the OPAMP is very high therefore the capacitor CH does not discharge and this is the HOLD phase for the S/H circuit. The sampling period is the length of time during which input voltage samples are created. During the holding phase, the OPAMP's output is processed. Thus, holding period is important for OP-AMPs. S/H circuit is used in Signal Constructional Filters, Sampling Oscilloscopes, Analog Signal Processing, Data Distribution System, Data Conversion System, Digital Voltmeters etc.

It must be noted that the input data is kept constant for the sampling time when C=1 in positive pulse.

6.4.2 Quantization and Encoding

Quantizing and Encoding is used after S/H circuits in ADCs. Quantization is the procedure by which the input signal is matched with the appropriate discrete quantum after the reference signal has been divided into multiple quantization levels. While in Encoding each quantum (quantization levels) will be given a special digital code which will then be used to identify the input signal level. Figure 6.8 provides an example of quantization noise and quantizing levels and corresponding encoding.

It can be noted here that the entire range of voltage in an interval is represented by a single digital value. As a result, there will be an error, known as a quantization error (σ_e) permitted up to ±0.5 LSB as is clear from Figure 6.8(a). This is the noise that the quantization procedure introduced. The quantization levels are 2N for N –bit resolution ADC and step size (Δ)= $V_{ref}/2^N$.

For example in 3-bit resolution there will be 2_3 = 8 quantization levels with step size (Δ) = $V_{ref}/23$ (=10/8=1.25 V for below graph)voltage difference for every next code in sequence. While in 16-bit ADC total quantization levels are 2^{16} with step size (Δ) = $V_{ref}/2^{16}$ voltage difference between levels. For Fig. 6.8(b), Δ =$V_{ref}/2^{16}$ = 152 μV voltage for each level. It can also be noted from the below graphs that more the resolution more the accuracy for ADC. Uniform power density function (PDF) for quantization noise introduced is given by $\sigma_e^{\,2} = \dfrac{\Delta^2}{12}$.

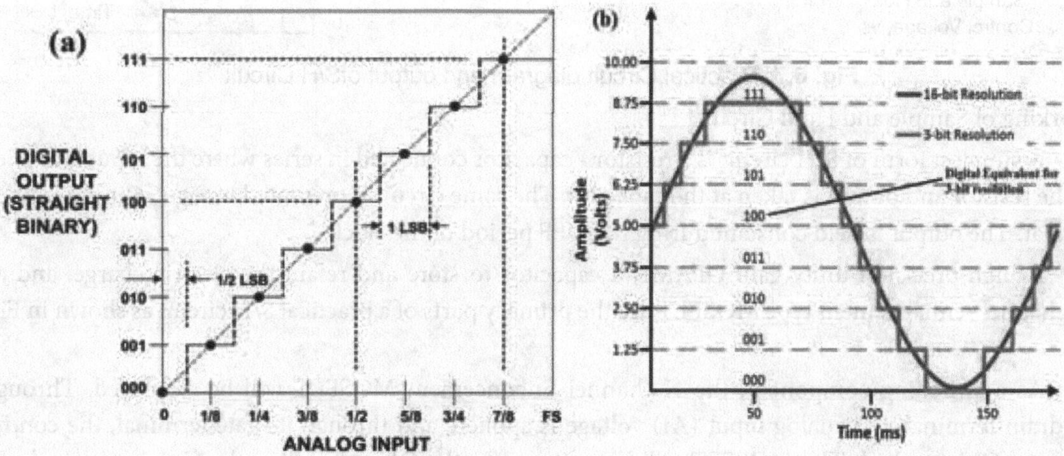

Fig. 6.8: (a) Quantization noise (b) Quantization levels and encoding.

6.5 SPECIFICATIONS FOR ADC

The main performance parameters for an ADC converter are discussed in this section.

6.5.1 Resolution

An ADC's resolution is the amount of analog input voltage change necessary to advance its digital output from one code to the next higher code. One component can be resolved by an N-bit ADC converter in 2N. It can be specified in bits or as a percentage of the full scale range (FSR) of voltage. For instance, one part in 256, 0.4% of FSR, or simply 8-bits resolution can be used to describe the resolution of an eight-bit ADC converter.

6.5.2 Gain and Offset Errors

The discrepancy between the actual and ideal FSR voltages is known as the gain error. Both LSBs and a percentage FSR are used to express it. When an ADC converter is working in bipolar mode, the offset error is the error at analog zero. It is quantified in LSBs or in percentage of FSR.

6.5.3 Gain and Offset Drifts

The variation in the FSR transition voltage over the course of the whole operational temperature range is known as gain drift. For an ADC converter operating in bipolar mode, the offset drift is the variation in the analog zero caused by temperature. Typically, it is stated as ppm of FSR or LSBs per degree Celsius.

6.5.4 Accuracy

The accuracy s specifies quantization error of the ADC converter, mostly from the comparator and ladder resistors as shown in Fig. 6.9. Gain, offset, and quantization problems make up the majority of these mistakes. Actual analog input and the ideal equivalent of the output code that corresponds to actual analog input are both described by accuracy.

6.5.5 Sampling Frequency and Aliasing Phenomenon

According to sampling theorem Nyquist criteria the analog signal is sampled at least twice the baseband frequency of input signal (Nyquist criteria). The quantization error, however, imposes a limit on the signal reproduction's accuracy. The replication of the signal is not an accurate representation of the original signal if the sampling rate is violating Nyquist criteria, leading to the formation of these fictitious signals, also known as aliases. The analog input passed through low-pass filter (LPF) to eliminate all frequency components above half the sampling rate in order to prevent the aliasing issue. All practical ADC converters employ this analog filter, known as an anti-aliasing filter.

6.5.6 Quantization Error

In section 6.4.2, quantization error is discussed. It can be understood by the below Fig. 6.9. this is the error between original analog input signal and the recreated analog input signal from ADC-DAC conversion. For N-bit ADC step size (Δ) = Vref/2N and uniform power density function (PDF) for quantization noise introduced is given by $\sigma_e^{\ 2} = \dfrac{\Delta^2}{12}$.

6.5.7 Integral Nonlinearity

Integral non-linearity (INL) indicates how actual output deviates from a linear transfer curve. Gain, offset, and quantization mistakes are not considered nonlinearity errors. It can be expressed in LSBs or as a percentage of FSR.

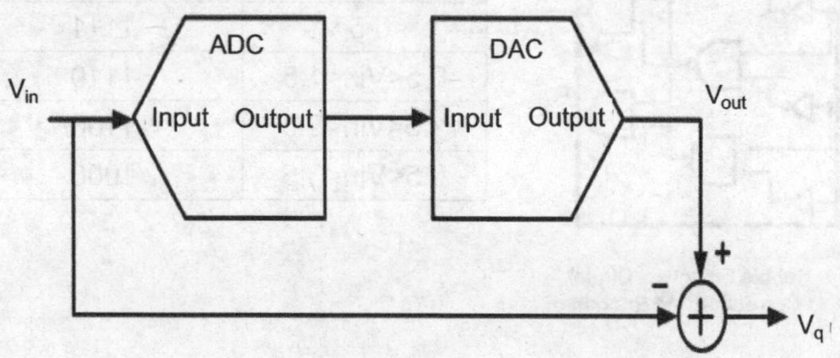

Fig. 6.9: Quantization error voltage output.

6.5.8 Conversion Time

This is the total time that passes between the instant the conversion signal starts and the moment the conversion complete signal appears. It varies from milliseconds to nanoseconds for different ADCs.

6.5.9 Aperture and Acquisition Times

The aperture time is the amount of time needed for the switch (MOSFET) to ON after the hold signal in S/H circuit is present. The time needed for the switch (MOSFET) to OFF and retain the signal in CH in form of in S/H circuit is called acquisition time. The ideal value for both times is 0. Thus, along with the conversion time, the aperture and acquisition periods also affect the maximum sampling frequency.

6.6 TYPES OF ADC CONVERTER

There are various ADCs, which vary in the speed of operation, accuracy and complexity. A few of these are explained here. ADCs are having a huge market size and a lot of applications in various domains. Broadly, there are two types of ADCs-

1. **Voltage domain ADCs** such as- FLASH ADC, SAR ADC, pipeline ADC. There are certain problems with voltage domain ADCs which can be overcome by time domain ADCs.
   (a) Transistors are optimized from digital design's perspective
   (b) Threshold voltage does not scale linearly with power supply
   (c) State-of-the-art processes have smaller gain transistors
   (d) High speed and resolution is usually achieved using pipelining
   (e) Pipelined ADC performance is dependent on OPAMP
2. **Time domain ADCs** such as -Dual slope ADC,V to F ADC, V to T ADC etc., can overcome the above listed many issues.

6.6.1 FLASH ADC

Flash ADC circuit, which is also known as the Parallel Comparator ADC, is the easiest and fastest ADC of voltage domain type ADC. It consists of a resistive voltage divider to set corresponding

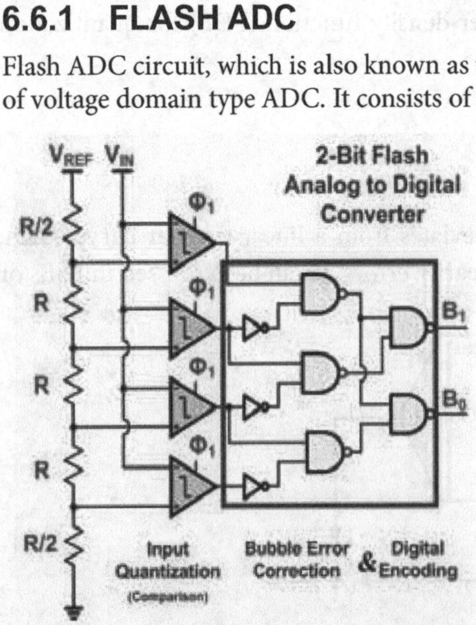

Working for Flash ADC

Input(Vin)	Encoder input ($Y_3Y_2Y_1Y_0$)	Encoder output(B_1B_0)
<0.5 V	1111	"00"
0.5<Vin<1.5	1110	"01"
1.5<Vin<2.5	1100	"10"
2.5<Vin<3.5	1000	"11"

Fig. 6.10: Two bit Flash ADC and it's working assuming Vref = 4 V.

reference level followed by a set of comparators, each of which assesses the input signal against a different reference voltage level as shown in Fig. 6.10. The priority encoder circuit, which generates a binary output, is connected to the comparator outputs at its inputs. An example of a 2-bit flash ADC and its working using priority encoder circuit is shown in Fig. 6.10. Here R/2 resistances are taken to give the rage of reference levels from $V_{LSB}/2$ to V_{ref} - $V_{LSB}/2$.

If instead of two R/2 resistances, "R" resistance is taken at the bottom only this will lead the range of reference levels from 0 to Vref - VLSB only which is not giving the scope for "-0.5 LSB" error for "0" reference level. The Priority encoder inputs are taken active LOW and outputs are active HIGH as shown in Table listed in Fig. 6.10.

The Flash ADC circuit includes a precision voltage regulator that supplies Vref, a stable reference voltage that is not included in the schematic. The comparator outputs will successively become saturated to a high state as the analog input voltage rises over the reference level value at each comparator. The highest-order active input is used by the priority encoder to generate a binary number while disregarding all other active inputs.

The speed of Flash ADC is only constrained by gate propagation delays, comparator & latches. 2N comparators are needed for this N-bit flash ADC. It would take 8, 16 comparators for a three/four-bit Flash ADC version respectively. The flash converter also has a non-linear output feature, which is a benefit that is frequently missed. Increasing number of bits makes the Flash ADC more complex circuit.

Drawbacks of Flash ADC-

- The exponential growth of comparators & resistors makes it "costly" as a function of resolution.
- Bubble error
- Less accuracy
- Cost is increased due to increment in the following-
    - ➢ comparator kickback noise – which is unwanted
    - ➢ power consumption – more number of comparators lead to consume more power
    - ➢ input capacitance- Increases conversion time and loading effect
    - ➢ chip area and difficulties in routing the signals

6.6.2 Successive Approximation (SAR) ADC

The successive approximation type ADC converter only attempts one bit at a time to approximate the analog signal that will be converted to digital data. It is a voltage domain type ADC. A very unique counter circuit called a successive-approximation register (SAR) is the only modification to this design. The block diagram for SAR ADC is shown in Fig. 6.11. This SAR register counts by testing all bit values starting with the MSB and ending with the LSB, it decides only "1" bit in one cycle. The SAR compares Vin with VDC in ith cycle to decide the ith bit as per Fig. 6.12. The register counts in the same way as the "trial-and-fit" method of converting numbers from decimal to binary, which involves attempting various bit values from MSB to LSB until a digital binary word is obtained that is equal to the initial analog input. The advantage of this counting method is that it yields results much more quickly like binary search algorithm.

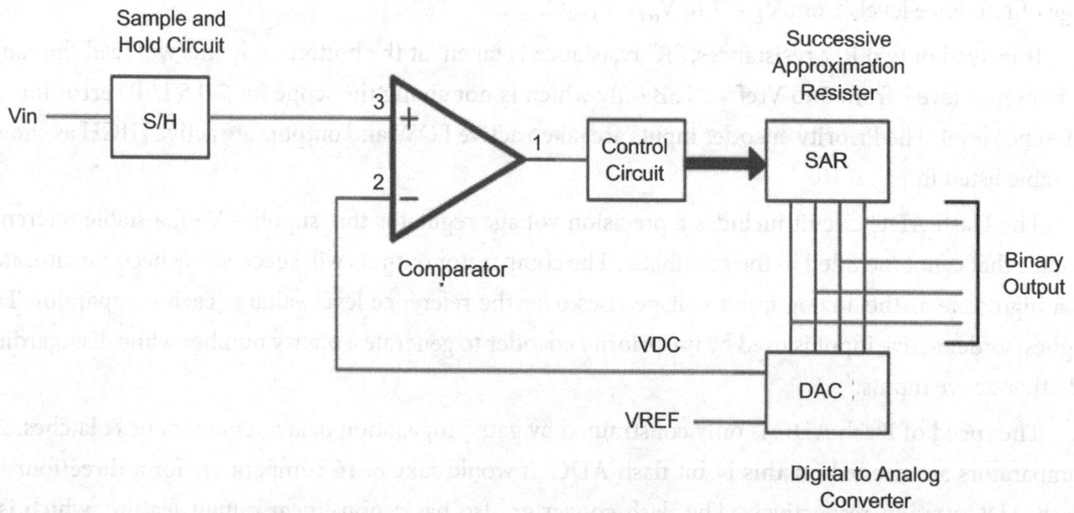

Fig. 6.11: Successive Approximation ADC Converter.

Note: that the SAR typically has the ability to output binary numbers in serial format, negating the requirement for a shift register.

SAR ADC Working

The algorithm for SAR ADC can be summarised in the following steps (assume 4-bit ADC), also explained in Fig. 6.12.

1. Set MSB =1 and rest all to "0" thus 1000 is the input bit set which has corresponding voltage VDC =Vref/2 obtained as output of DAC. (This is equivalent to divide complete voltage rage in two parts)
2. Compare Vin with VDC- (assume b1 is MSB)
   (a) If Vin > V_{DC} Keep this bit bi= 1 and $V_{DC} = V_{DC} + V_{ref}/2^{i+1}$
   (b) Else reset bi = 0 and $V_{DC} = V_{DC} - V_{ref}/2i + 1$
3. Assume next b^{i+1} =1 and repeat step 2 for all bits.
4. The bit set obtained finally is the output of ADC

Advantages of SAR ADC	Disadvantages of SAR ADC
Moderate speed of operation but better accuracy	Design complexity
Low power	Higher cost

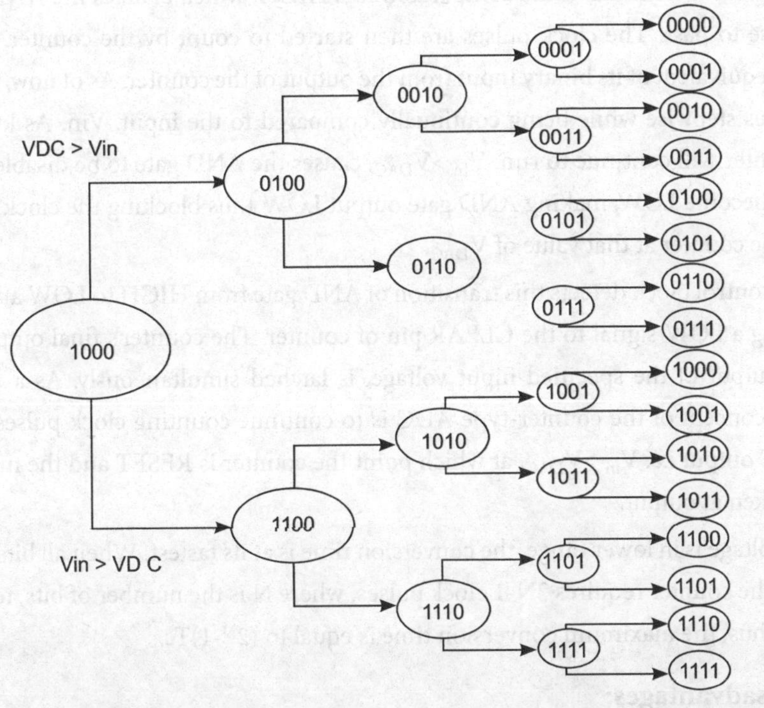

Fig. 6.12: Successive Approximation ADC Converter working.

SAR ADC Applications-

SAR ADC is most widely used due to its better speed and accuracy. It is used in sensors, medical applications to be implanted in patients due to low power, smart watches etc. Its general speed range is 2-5 mega samples per second (MSPS) & found in resolution of 8-16 bits generally.

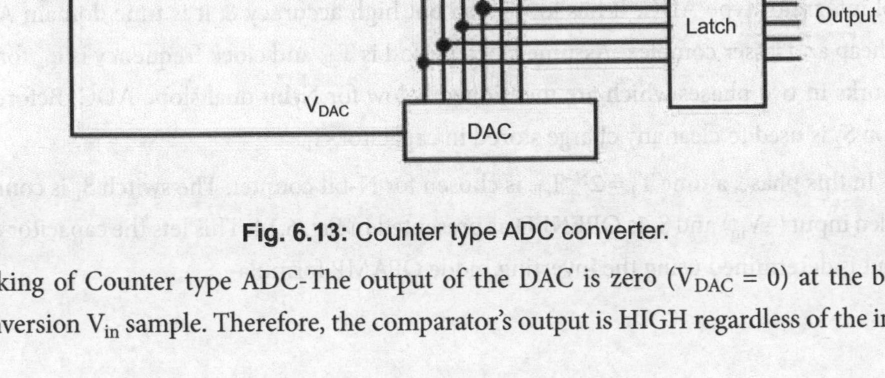

Fig. 6.13: Counter type ADC converter.

Working of Counter type ADC-The output of the DAC is zero ($V_{DAC} = 0$) at the beginning of every conversion V_{in} sample. Therefore, the comparator's output is HIGH regardless of the input voltage

(V_{in}) provided to the positive terminal of the comparator. It is HIGH, which enables the AND gate and permits the clock pulse to pass. The clock pulses are then started to count by the counter. The DAC computes the decimal equivalent of its binary input from the output of the counter. As of now, the output of the DAC, V_{DAC}, rises stepwise while being continually compared to the input, Vin. As long as Vin exceeds V_{DAC}, the counter will continue to run. $V_{in} > V_{DAC}$, causes the AND gate to be disabled and the comparator output to become LOW, making AND gate output LOW thus blocking the clock pulses for counter which stops the counter at that value of V_{DAC}.

Additionally, the control block detects this transition of AND gate from HIGH to LOW and RESETs the counter by applying a LOW signal to the CLEAR pin of counter. The counter's final output, which is the digital binary output of the specified input voltage, is latched simultaneously. As a result, the fundamental working concept of the counter-type ADC is to continue counting clock pulses until the input exceeds the DAC output i.e. $V_{in} > V_{DAC}$, at which point the counter is RESET and the most recent count is latched and taken at output.

When the input voltage is in lower range, the conversion time is at its fastest. When all bits are 1, the DAC outputs at FSR. The counter requires 2N-1 clock pulses, where N is the number of bits, to reach all ones from all zeroes. Thus, the maximum conversion time is equal to (2^N-1)Tc.

Advantages and Disadvantages:

- The counter-type ADC is straightforward and user-friendly. It is extremely precise, and adding more bits will increase the precision. Despite the market's abundance of complex ADCs, the counter-type ADC strikes a good compromise between precise output and affordable hardware prices.
- This ADC's primary flaw is that each time a new conversion begins, the counter is RESET and must begin counting from zero. The conversion time is important as a result. As it won't be able to convert the analog signal to digital binary in real time, it cannot withstand high-frequency input. When the input voltage is equivalent to the DAC's full-scale output range, the worst case scenario happens.

6.6.4 Dual Slope ADC Converter

It is a type of integrator type ADC. It has low speed but high accuracy & it is time domain ADC. It is reasonably cheap and lesser complex. Assume clock period is T_{clk} and clock frequency is f_{clk} for counter. This ADC works in two phases which are mentioned below for N-bit dual slope ADC. Before starting the conversion S_2 is used to clear any charge stored in capacitor C_1.

Phase I- In this phase, a time $T_1 = 2^N * T_{clk}$ is chosen for N-bit counter. The switch S_1 is connected to analog sampled input ($-V_{in}$) and S_2 is OPENED as presented in Fig. 6.14. This lets the capacitor to charge and the output is determined using the inverting mode OPAMP formula-

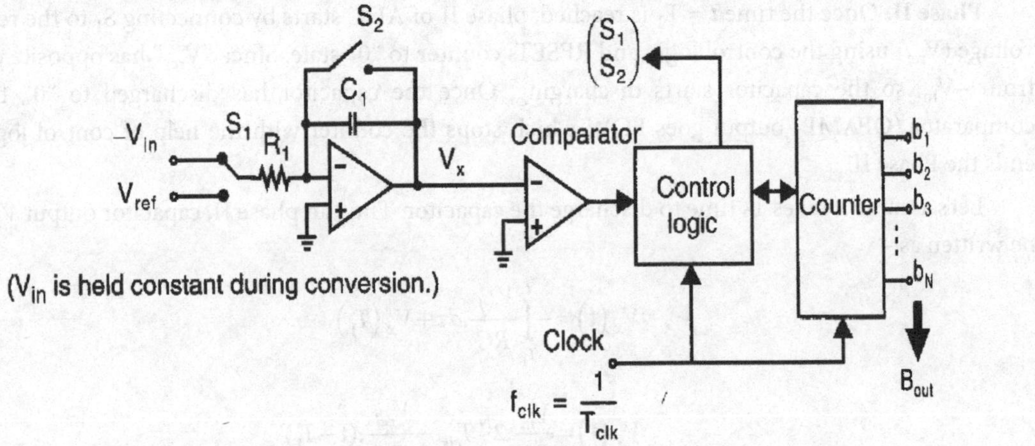

Fig. 6.14: Block schematic representation of a dual-slope ADC converter.

$$V_x(t) = -\int_0^t \frac{(-V_{in})}{RC}.dt = \frac{V_{in}}{RC}.t$$

This is an equation of line with variable slope $\frac{V_{in}}{RC}$. Thus, phase I, has variable slope of line which changes as per V_{in} variations can be understood from Fig. 6.15.

Since this $V_x(t) < 0$ which in turn gives HIGH output for 2nd OPAMP & the control logic starts the counter. Thus counter starts counting from "0" at $t = 0$.

The maximum output voltage when $t = T_1$ is given as-

$$V_x(Max) = \frac{V_{in}}{RC}.T_1 = \frac{V_{in}}{RC}.2^N T_{clk}$$

And counter has counted corresponding 2^N states in this duration.

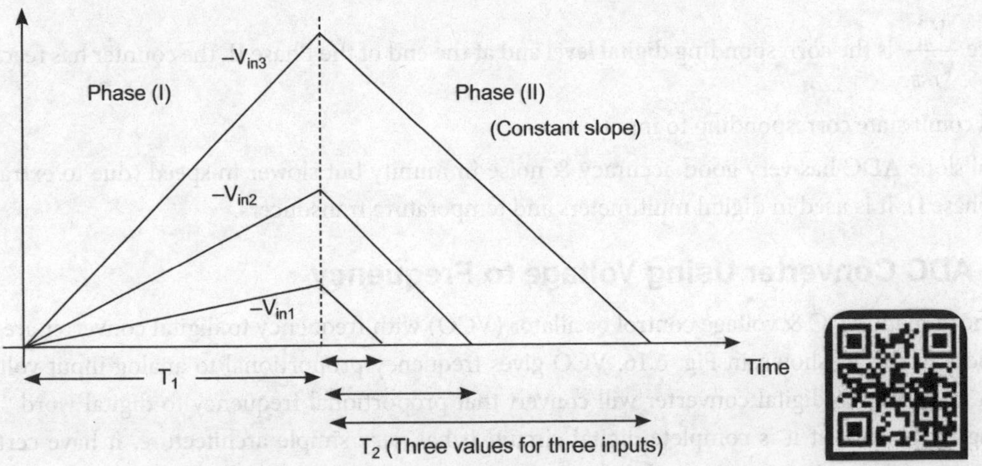

Fig. 6.15: Operation of Dual-slope ADC for different input voltages.

Phase II. Once the time $t = T_1$ is reached, phase II of ADC starts by connecting S_1 to the reference voltage (V_{ref}) using the control logic and RESETs counter to "0" state. Since "V_{ref}" has opposite polarity from $-V_{in}$, so the capacitor starts discharging. Once the capacitor has discharged to "0", the 2nd comparator (OPAMP) output goes LOW which stops the counter with the help of control logic, this ends the Phase II.

Let's assume it takes T_2 time to discharge the capacitor. Thus, in phase II, capacitor output $V_x(t)$ can be written as -

$$V_x(t) = -\int_{T_1}^{t} \frac{V_{ref}}{RC} . d\tau + V_x(T_1)$$

$$V_x(t) = \frac{V_{in}}{RC} .2^N T_{clk} - \frac{V_{ref}}{RC} .(t - T_1)$$

Here it is to be noted that the phase II line slope is constant $(-\frac{V_{ref}}{RC})$. Therefore, phase II is also called constant phase.

Since $V_x(t) = 0$, is the end of phase II, when $t = T_1 + T_2$ time. Thus the above equation can be written for $t = T_1 + T_2$ as –

$$0 = \frac{V_{in}}{RC} .2^N T_{clk} - \frac{V_{ref}}{RC} .T_2$$

$$\Rightarrow \quad T_2 = \frac{V_{in}}{V_{ref}} .2^N T_{clk} = \frac{V_{in}}{V_{LSB}} .T_{clk}$$

Or

$$\frac{T_2}{T_{clk}} = \frac{V_{in}}{V_{LSB}}$$

Since $\frac{V_{in}}{V_{LSB}}$ is the corresponding digital level and at the end of the Phase II, the counter has reached to digital count state corresponding to input.

Dual slope ADC has very good accuracy & noise immunity but slower in speed (due to extra T1 time in phase I). It is used in digital multimeters and temperature transducers.

6.6.5 ADC Converter Using Voltage to Frequency

It is a time domain ADC & voltage control oscillator (VCO) with frequency to digital converter are the main blocks in this as shown in Fig. 6.16. VCO gives frequency proportional to analog input voltage and then frequency to digital converter will convert that proportional frequency to digital word. The best thing in this is that it is complete digital circuit. It has very simple architecture. It have certain disadvantages- (a) the VCO linearity is hard to maintain (b) range of VCO frequencies is limited (c) to improve signal to noise ratio (SNR), it requires post processing.

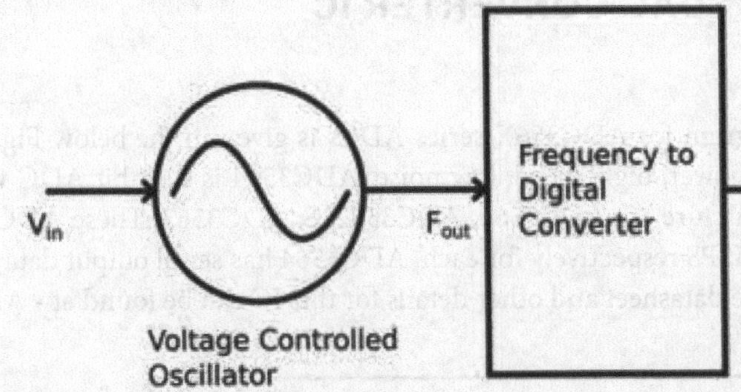

Fig. 6.16: Voltage to Frequency converter ADC .

6.6.5 ADC Converter Using Voltage to Time

The fundamental voltage-to-time conversion type of an ADC is depicted in Fig. 6.17. It has two blocks- (a) voltage to time converter (VTC) & (b) time to digital converter (TDC). The VTC accepts the analog sampled input and converts this to time (it can use an integrator as shown in dual slope ADC). Once this conversion is done the TDC converts that time to digital word.

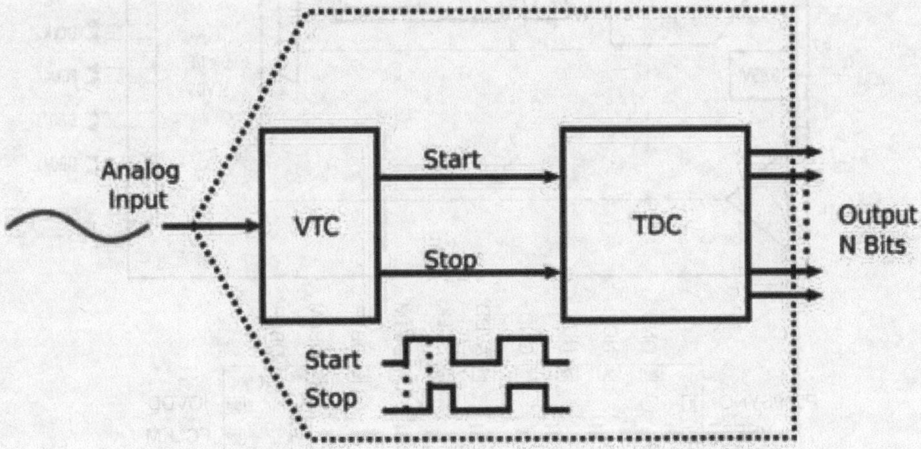

Fig. 6.17: ADC converter using voltage to time converter.

Advantages of V to T ADC:-

1. Analog and digital sections are separated to better noise immunity
2. Analog section can be made lesser complex
3. Low power high speed ADC can be designed

Disadvantages of V to T ADC:-

1. VTC is highly sensitive, so design carefully for linearity and noise immunity perspectives.
2. Hybrid technique must be employed for higher linearity.

6.7 EXAMPLE OF ADC & DAC CONVERTER IC

6.7.1 ADC3564

The basic block diagram & pin diagram for ADC356X series ADCs is given in the below Fig. 6.18. It is a 40 pin IC which has ultra-low power, high speed, low noise. ADC3564 is a 14-bit ADC with 125 MSPS speed. ADCs with similar structure are ADC3561, ADC3562, & ADC3563. These ADCs are 16 bit ADCs and speed is 10, 25 & 65 MSPS respectively for each. ADC3564 has serial output data and can work at a wide temperature rage. The datasheet and other details for this IC can be found at - ADC3564 datasheet | TI.com

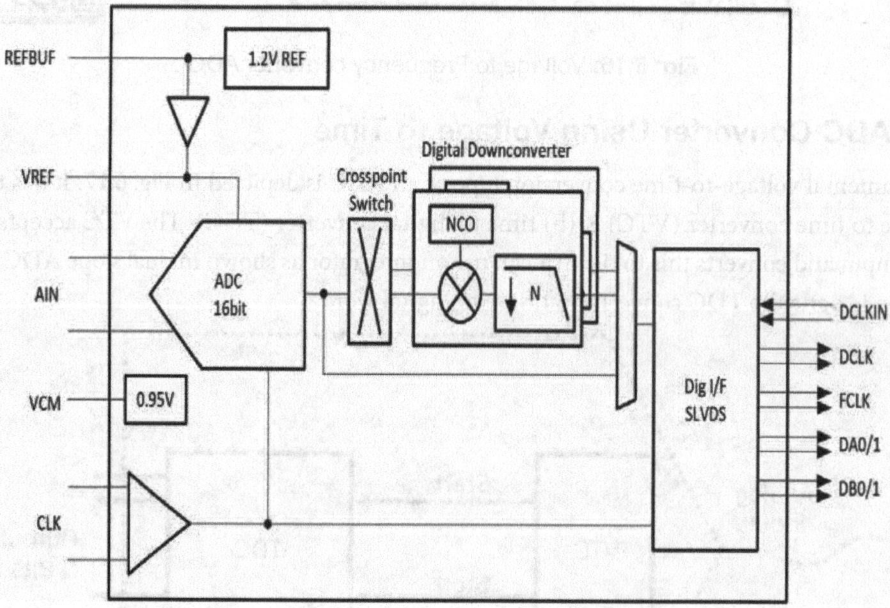

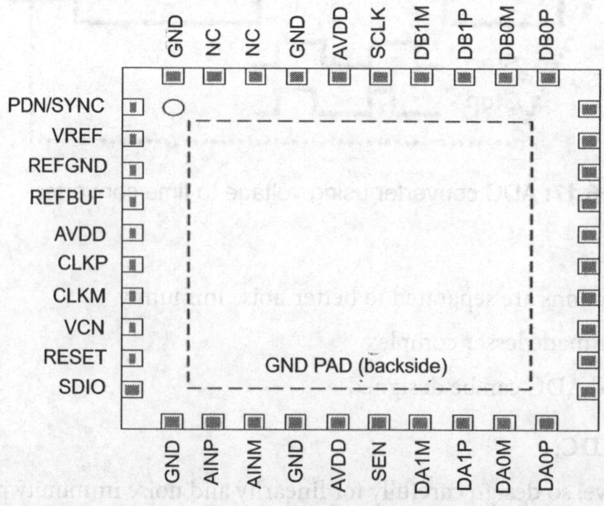

Fig. 6.18: Block diagram for ADC356X series by TI & pin diagram for ADC3564.

Applications- ADC3564 has typical applications in – spectroscopy, smart grid, radar, thermal imaging etc.

6.7.2 SAR ADC MAX19777

This is SAR ADC provided by Maxim Integrated Company. MAX19777 ADC is dual channel and uses 2X1 MUX to select any one channel. It is a low power 8-pin ADC available in wafer level packaging (WLP). It has very good figures of merit. It can be used conveniently. It has feature set like – speed is 3 MSPS, wide temperature range, serial peripheral interface, low power consumption. More details about this IC can be found at - MAX19777-3Msps, Low-Power, Serial 12-Bit ADC (maximintegrated.com).

Applications- medical, communication, data acquisition, portable data logging etc.

Other few examples of ADC ICs are - AD ADC 800, AD ADC 80, AD ADC 84 etc.

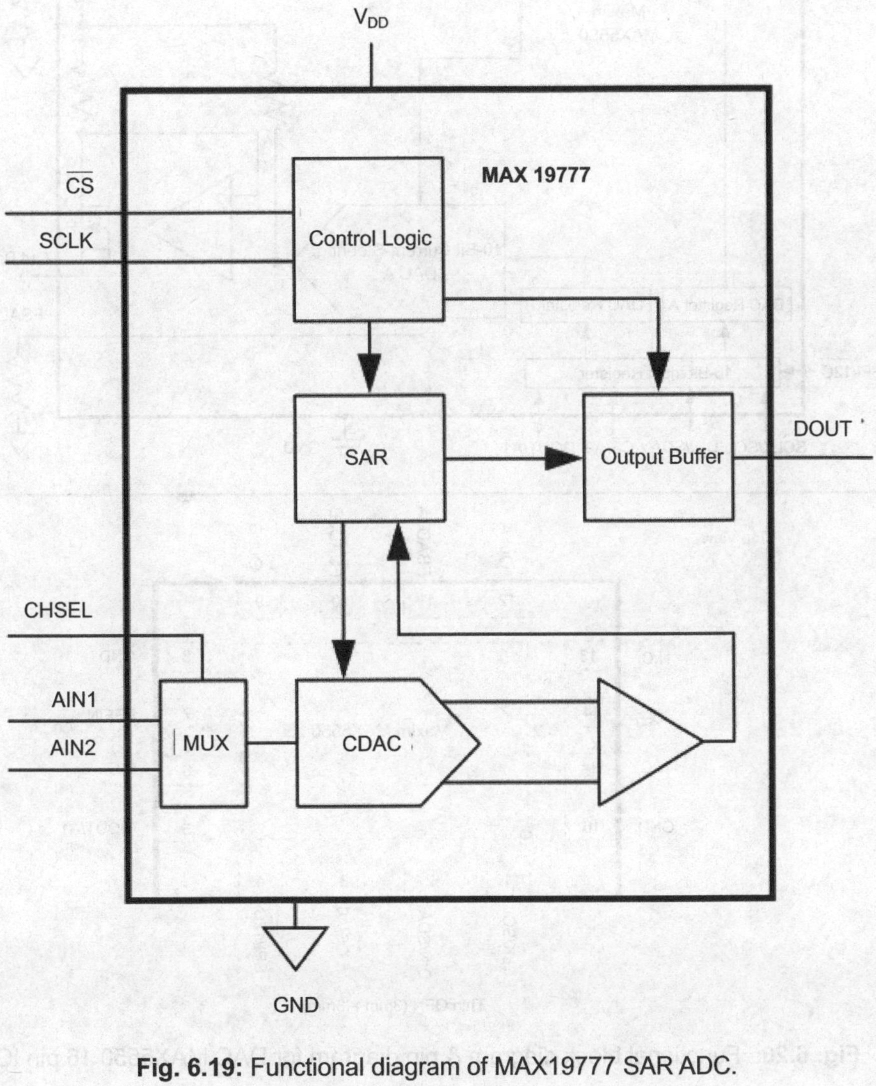

Fig. 6.19: Functional diagram of MAX19777 SAR ADC.

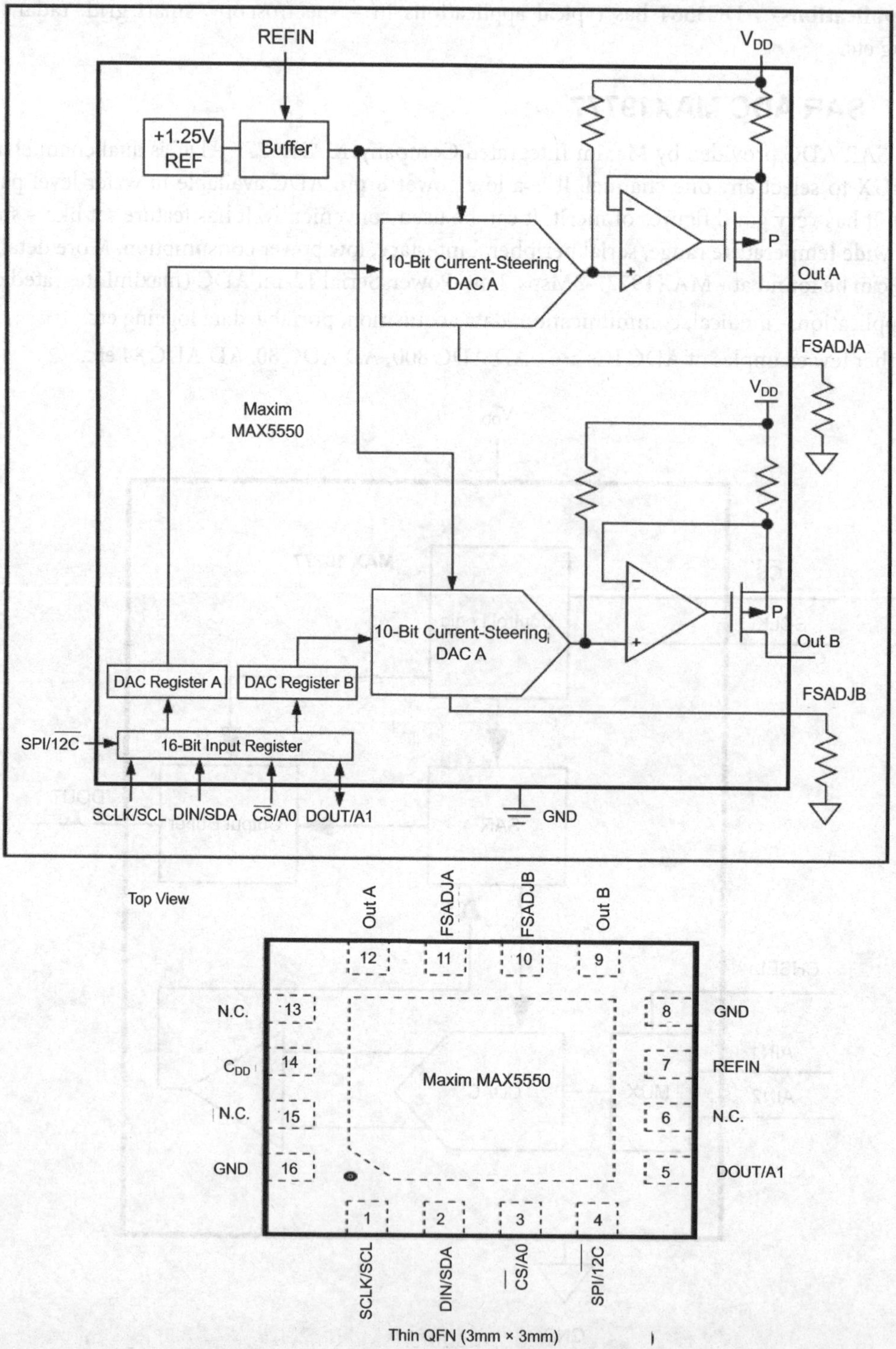

Fig. 6.20: Functional block diagram & pin diagram for DAC MAX5550 16 pin IC.

6.7.3 DAC MAX5550

This is a 10 bit dual channel DAC with high output current carrying capability. So this DAC IC can be used for high output current applications. It provides serial interfacing. It is 16 pin IC and monotonic over the complete operating temperature ranges. Its output current can be programmed to change. It's functional block diagram & pin diagram are presented in Fig. 6.20. A more details about this IC can be found at - MAX5550 DS (maximintegrated.com)

Applications-VCO tuning, RF attenuator control.

6.7.4 DAC 712

DAC712 is a 28-pin dual-in-package (DIP) DAC IC provided by TI company, which is completely monotonic over complete temperature range with 16-bit resolution. It is a high speed DAC with output range ±10 V. DAC712 also has the facility to adjust the gain and offset error. Fig. 6.21 shows functional and pin diagrams for DAC712. More details about it can be found at - https://bit.ly/3EFjXHm or scan the below QR code for more details.

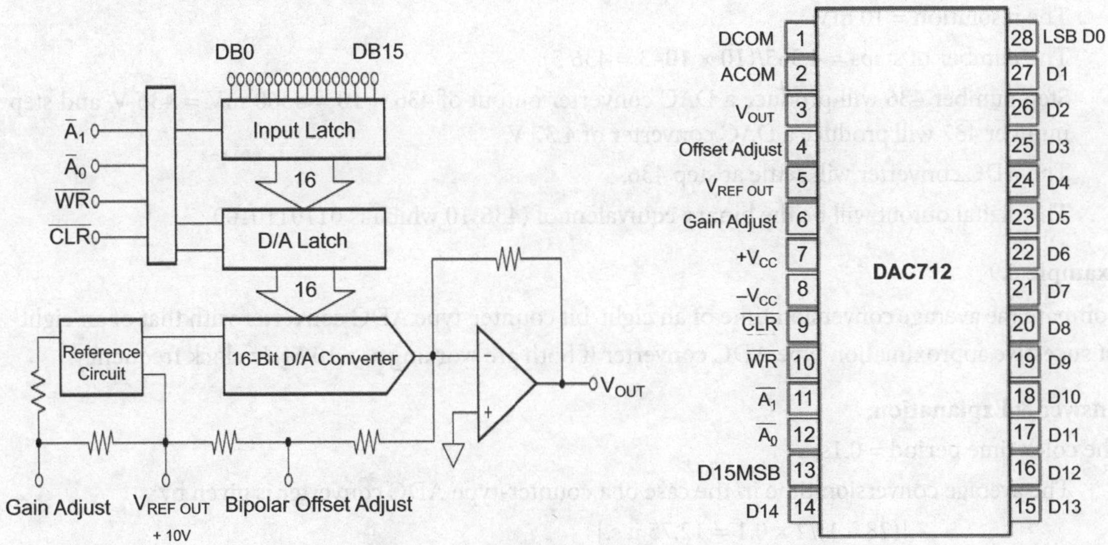

Fig. 6.21: Functional block diagram & pin diagram for DAC712 -16 pin IC.

Example 6.7

A specific 8-bit ADC with a full-scale input of 2.55 V (V_A = 2.55 V yields an output of 11111111 on the digial scale). It has a 0.1% F.S. defined error. Calculate the largest possible difference between the analogue input and V_{AX} output.

Answer & Explanation:

The step size is 2.55 V/(2^8 – 1), which is exactly 10 mV. This means that even if the DAC has no inaccuracies, the VAX output could be off by as much as 10 mV because VAX. can change only in 10-mV steps; this is the quantization error. The specified error of 0.1% × 2.55 V = 2.55 mV. This means that the VAX value can be off by as much as 2.55 mV because of component inaccuracies. Thus, the total possible error could be as much as 10 mV + 2.55mV = 12.55 mV.

Example 6.8

A 10-bit ADC converter of the successive approximation type has a resolution (or quantization error) of 10 mV. Determine the digital output for an analogue input of 4.365 V.

Answer & Explanation:

In the case of successive approximation type ADC converter the final analogue output of its DAC converter portion always settles at a value below the analogue input voltage to be digitized within the resolution of the converter.

- The analogue input voltage = 4.365 V.
- The resolution = 10 mV.
- The number of steps = 4.365/(10 × 10–3 = 436.5)
- Step number 436 will produce a DAC converter output of 436 × 10 = 4360 mV = 436 V, and step number 437 will produce a DAC converter of 4.37 V.
- The ADC converter will settle at step 436.
- The digital outout will be the binary equivalent of (436)10 which is 0110110100.

Example 6.9

Compare the average conversion time of an eight-bit counter-type ADC converter with that of an eight-bit sucessive approximation type ADC converter if both are working at a 10 MHz clock frequency.

Answer & Explanation:

The colck time period = 0.1s.

- The average conversion time in the case of a counter-type ADC converter is given by

  [(28 – 1)/2 × 0.1 = 12.75 μ.s.]
- The conversion time in the case of a successive approximation type ADC converter is given by 8 × 0.1 = 0.8 μ s.

UNIT SUMMARY

In this unit, digital to analog (DAC) and analog to digital (ADC) which are a necessary components for any interface between digital and analog devices. The operational basics, the key performance criteria and their significance, as well as the many forms and uses of digital-to-analog and analog-to-digital converters are covered in this chapter.

— DAC is used to get analog output for machine human interface.

— ADC is used to get digital output for human machine interface.

— R-2R ladder DAC can be designed just using two resistances.
— Number of bit in DAC or ADC increases the complexity of that converter circuit.
— Dual slope ADC is much accurate but slower in speed. It uses two phases to complete the conversion.
— SAR ADC decides one bit in one cycle and works like binary search algorithm
— SAR ADC is with medium speed and medium accuracy.
— Flash ADC is the fastest and the simplest one. Flashback error, bubble error and matching of comparators are difficulties to design Flash ADC.
— There are commercial ICs available for ADCs and DACs by companies like TI, Maxim Integrated etc. These ICs can be easily used for experimental and other purposes.

EXERCISES

Multiple Choice Questions

1. In addition to the reference voltage, how many control lines are there in an analog to digital converter.
   (a) 3 (b) 2
   (c) 1 (d) None
2. Determine the type of integrating analog to digital converter?
   (a) Counter type (b) Dual slope ADC
   (c) Flash type (d) Tracking
3. Which ADC converter is the cheapest, fastest, and most straightforward?
   (a) Servo type (b) Counter type
   (c) Flash type (d) All
4. What benefit does employing a flash type ADC converter offer?
   (a) Average conversion (b) Slow conversion
   (c) Fast conversion (d) None
5. Determine the time required to convert a full scale input using a 12-bit counter type ADC with a 1 MHz clock frequency.
   (a) 4.095 ms (b) 4.095 μs
   (c) 4.095 s (d) None
6. What circumstance causes a servo tracking ADC converter error?
   (a) Average i/p conversion (b) Slow i/p conversion
   (c) Fast i/p conversion (d) None
7. Which application uses a dual slop converter.
   (a) Thermocouple (b) Digital panel meter
   (c) Weighting scale (d) All
8. The highest output voltage for a dual ramp generator with 12 bits is +12v. For the analog signal of +6v, calculate the corresponding digital number.
   (a) 1000000000 (b) 10000000000
   (c) 100000000000 (d) 1000000000000

9. When this occurs, the input voltage of a servo tracking ADC converter is greater than the DAC output signal.
   (a) Count up (b) Count down
   (c) Back and forth (d) None
10. How many clock pulses are needed to produce a digital output by a consecutive approximation converter.
   (a) 12 (b) 6
   (c) 8 (d) None

ANSWERS															
1.	(b)	2.	(b)	3.	(c)	4.	(c)	5.	(a)	6.	(c)	7.	(b)	8.	(c)
9.	(a)	10.	(d)												

Short and Long Answer Type Questions

1. Give a brief explanation of the binary ladder network's digital-to-analog conversion mechanism. What distinguishes it from the equivalent basic resistive network?
2. Distinguish between resolution and accuracy in relation to ADC converters, as well as nonlinearity (NL) and differential nonlinearity (DNL)?
3. Using a schematic diagram, explain the sequential approximation type ADC converter's working principles?
4. Why fast-changing analog signals are especially well-suited for tracking type ADC converters?
5. Why the accuracy of a dual-slope integrating-type ADC converter is greater than that of a single-slope integrating-type ADC converter?
6. Explain any use of a DAC as a programmable integrator?
7. What are the advantages and disadvantages of Flash type converter?
8. What is a multiplying-type DAC converter?
9. Discuss any one of the DAC IC with internal details.
10. Discuss any one of the ADC IC with internal details.

Numerical Problems

1. Find the percentage resolution of an 8-bit and a 12-bit DAC converter, respectively.
2. The resolution of a 12-bit DAC converter is 2.44 mV. For a digital input of 11111000111, determine its analog output if all "0" is having 0 V output. Find Vref as well.
3. The full-scale analog input of one 12-bit successive approximation type ADC converter is 10 V. It uses a 1MHz clock frequency to run. Calculate the duration of the conversion for an analog input of -
   (a) 1.2 V, (b) 2.350 V, (c) 4.75 V
   What happens if – i) the clock frequency is 2 MHz ii) bits increased to 16-bit ADC
4. For a current-output DAC converter to have a full-scale output of 20 mA and a resolution better than 25 mA, how many bits need it have?
5. For a digital input of 00000010, an eight-bit DAC converter generates an analog output of 12.5 mV. Find the analog output for an input of the kind 0000100 in digital form.

PRACTICAL

Experiment 10. Study A to D and D to A Conversion Operation

Please refer UNIT VI of Digital Electronics book.

Experiment 11. Simulate in Software (such as multisim or PSpice) an Analog to Digital Converter

Please perform the experiment using below video link- https://youtu.be/4l_bf2WTAGQ

Experiment 12. Simulate in Software (such as multisim or PSpice) a Digital to Analog Converter

Please perform the experiment using below video link -https://youtu.be/zdMk2YpTmp0

KNOW MORE

- Weighted resistor DACs can have a bit different architecture to accommodate more resolution like folding architecture is the one which uses decoders lines to select the corresponding voltage level.
- Converters can work on the principle of charge sharing as well where a number of capacitors are taken instead of resisters for weighted binary DAC.
- Current sharing DAC converters are easy to design using thermometer code converters.

REFERENCES AND SUGGESTED READINGS

Johns, David A., and Ken Martin, "Analog integrated circuit design." John Wiley & Sons, 2008.

Dynamic QR Code for Further Reading

QR codes are put along with topics in the unit.

REFERENCES FOR FURTHER READING

List of some of the books is given below which may be used for further learning of the subject (both theory and practical):

1. M. Morris Mano and Michael Ciletti, "Digital Design", 5th edition, Pearson.
2. Johns, David A., and Ken Martin, "Analog integrated circuit design." John Wiley & Sons, 2008.
3. https://nptel.ac.in/courses/108105132
4. https://nptel.ac.in/courses/108105113
5. https://www.edaplayground.com

CO AND PO ATTAINMENT TABLE

Course outcomes (COs) for this course can be mapped with the programme outcomes (POs) after the completion of the course and a correlation can be made for the attainment of POs to analyse the gap. After proper analysis of the gap in the attainment of POs necessary measures can be taken to overcome the gaps.

Table for CO and PO attainment

Course Outcomes	Attainment of Programme Outcomes (1- Weak Correlation; 2- Medium Correlation; 3- Strong Correlation)						
	PO-1	PO-2	PO-3	PO-4	PO-5	PO-6	PO-7
CO-1	-	-	-	-	-	-	-
CO-2	-	-	-	-	-	-	-
CO-3	-	-	-	-	-	-	-
CO-4	-	-	-	-	-	-	-
CO-5	-	-	-	-	-	-	-
CO-6	-	-	-	-	-	-	-

The data filled in the above table can be used for gap analysis.

INDEX

□□